THE ORGANIZATIONAL ALIGNMENT HANDBOOK

A Catalyst for Performance Acceleration

THE ORGANIZATIONAL ALIGNMENT HANDBOOK

A Catalyst for Performance Acceleration

H. James Harrington • Frank Voehl

CRC Press
Taylor & Francis Group
Boca Raton London New York

CRC Press is an imprint of the
Taylor & Francis Group, an **informa** business
A PRODUCTIVITY PRESS BOOK

CRC Press
Taylor & Francis Group
6000 Broken Sound Parkway NW, Suite 300
Boca Raton, FL 33487-2742

© 2012 by Taylor & Francis Group, LLC
CRC Press is an imprint of Taylor & Francis Group, an Informa business

No claim to original U.S. Government works

ISBN-13: 978-1-4398-7732-6 (hbk)

Library of Congress Cataloging-in-Publication Data

Harrington, H. J. (H. James)
    The Organizational alignment handbook : a catalyst for performance acceleration / H. James Harrington, Frank Voehl.
        p. cm.
    Includes bibliographical references and index.
    ISBN 978-1-4398-7732-6 (hardcover : alk. paper)
    1. Strategic planning. 2. Organizational change. 3. Organizational effectiveness. 4. Reengineering (Management) 5. Industrial productivity. I. Voehl, Frank, 1946- II. Title.

HD30.28.H3733 2012
658.4'06--dc23   2011038960

Visit the Taylor & Francis Web site at
http://www.taylorandfrancis.com

and the CRC Press Web site at
http://www.crcpress.com

Dedication

I dedicate this book to my new granddaughter, Grace. I look

forward to many happy moments with her and her parents

over the years ahead. Work is good, but family is better.

—H. James Harrington

I dedicate this book to my two granddaughters, Abby Rose and

Emma Catherine, who both love to read and be read to, in spite

of their young ages. And to their two wonderful parents, Jeff

and Christina Voehl, who never met a book they didn't enjoy

and a story they didn't like. Perhaps my next book will be about

their love of books. Read on, my sweet ones, read on!

—Frank Voehl

Contents

The Management for Results Handbook Series

As series editors, we at CRC/Productivity Press have been privileged to contribute to the convergence of philosophy and the underlying principles of Management for Results, leading to a common set of assumptions. One of the most important deals with the challenges facing the transformation of the organization and suggests that managing for results can have a significant role in increasing and improving performance and strategic thinking, by drawing such experiences and insights from all parts of the organization and making them available to points of strategic management decision and action. As John Quincy Adams once said: "If your actions inspire others to dream more, learn more, do more, and become more, you are a leader."

If a good leader's actions inspire people to dream more, learn more, do more, and become more (and those actions will lead to an organization's culture, and if the culture represents "the way we do things around here") then the Management for Results Series represents a brief glimpse of "the shadow of the leader-manager." The series is a compilation of conceptual management framework and literature review on the latest concepts in management thinking, especially in the areas of accelerating performance and achieving rapid and long-lasting results. It examines some of the more recent as well as historical contributions, and identifies a number of the key elements involved. Further analysis determines a number of situations that can improve the results-oriented thinking capability in managers, and the various handbooks consider whether organizations can successfully adopt their content and conclusions to develop their managers and improve the business.

This is a particularly exciting and turbulent time in the field of management, both domestically and globally, and change may be viewed as either an opportunity or a threat. As such, the principles and practices of Management for Results can aid in this transformation or (by flawed implementation approaches) can bring an organization to its knees. This Management for Results Series (and the handbooks contained therein) discusses the relationship between management thinking, results orientation, management planning, and emergent strategy, and suggests that management thinking needs to be compressed and accelerated, as it is essential in making these relationships more appropriate and effective—a so-called "shadow of the leader-manager." As series editors, we believe that the greater the sum total of management thinking and thinkers in the organization, the more readily and effectively it can respond to and take advantage of the vast array of changes occurring in today's business environment. However, despite the significant levels of de-layering and flattening of structures that has taken place in the last decade or so, some organizational barriers continue to stifle opportunities for accelerating management for results by limiting the flow of experiences and insights to relevant corners of the organization.

The "shadow of the leader-manager" that is present throughout this management handbook series is based upon the following eight characteristics of an effective leader-manager who gets results, and provides one of the many integration frameworks around which this Series is based:

- Integrity = the integration of outward actions and inner values.
- Dedication = spending whatever time or energy necessary to accomplish the task at hand, thereby leading by example and inspiring others.
- Magnanimity = giving credit where it is due.
- Humility = acknowledging they are no better or worse than other members of the team.
- Openness = being able to listen to new ideas, even if they do not conform to the usual way of thinking.
- Creativity = the ability to think differently, to get outside of the box that constrains solutions.
- Fairness = dealing with others consistently and justly.
- Assertiveness = clearly stating what is expected so that there will be no misunderstandings and dealing with poor performance.

This management book series is intended to help you take a step back and look at your team or organization's culture to clearly see the reflection of your leader-manager style. The reflection you see may be a difficult thing for you to handle, but do not respond by trying to defend or to rationalize it as something not being of your making. As difficult as it may be, managers need to face the reality that their team and organization's culture is a reflection of their leadership, leading to the concept of the leader-manager.

Accepting this responsibility is the first step to change, and as we all know, change begins with ourselves.

As Ghandi said many years ago, we all need to strive to become the "change we want to see in the world" …. in the case of management for results, we need to be the change we want to see in others!

Frank Voehl and H. James Harrington

Series Editors

Preface

Too many organizations are organized to have the different functions compete with each other, not work together for the good of the total.

—HJH

MEMO TO THE CEO OF A FORTUNE 500 COMPANY

You have just left a board of directors' palaver where the members expressed their dissatisfaction with progress achieved last year. In your head you know that there must be a rational explanation for the failure of the organization to change; in your heart you feel that there are insurmountable obstacles. According to the latest employee surveys, the organization's culture or the group sense of motivation is the lowest that it has been in years. Unless you can demonstrate substantial change, your next task might be updating and faxing your résumé. You keep your résumé in a private compartment in the left-hand drawer of your desk. You can see yourself filling out the transmittal slip with the name of the headhunter you hired last year. You wonder if he will help, as the job he was handed to find a new senior vice president never turned up anyone that could help run the organization and the position is still vacant. Your future is on the line and so is the organization's.

During the past 20 years or so, many executives thought that quality management was a flag that everyone would salute. After some successful and not-so-successful launches and some interesting meetings with consultants, the initial enthusiasm has wavered and waned. There was little acceptance of this process by union employees. Many line managers have complained that there were few significant changes resulting from the process. As a leader, you vaguely remember a *Wall Street Journal* article describing the case where the CEO of one organization achieved the holy grail of total quality management (TQM)—the Malcolm Baldrige Award. Now he is looking for a job. Wasn't he one of the candidates for that vacant executive vice president slot and didn't the board turn him down? You are sure that was an omen.

The quality management process was informative although somewhat shallow, and you are now aware of several Lean Six Sigma approaches that Japanese executives use to reduce production defects. They also use the approaches to generate teamwork for solving manufacturing glitches and to iron out customer service snafus. None of these ideas has resulted in the catalyst for change that will help transform your organization. There have been no breakthrough strategies. The only person who knows what that means or thinks he does is the human resources vice president who thinks she deserves a medal for using the word "downsizing" at every meeting she attends.

Without a major breakthrough, your competitor's market share will continue to expand. With Six Sigma, the rate of decline is slower this year but as one board member said, that just prolongs the day when our once-dominant market share numbers convert into second place. The impact of even the slow rate of decline has been horrendous. Your stock options are now almost worthless.

Imitating the Japanese has proven to be an elusive goal. You want to leapfrog competition, not emulate them. You have observed that the institutional mechanisms in Japan are not the same as in your country. Given that very interesting visit to Nagoya last year, you know that it may take years to fix all of the cultural differences that exist. You may never have workers who volunteer their spare time to generate 2000 implemented suggestions a year as they do at Toyota. Values take generations to change, and you really only have the next 12 months, at best.

You also participated in cross-functional quality team meetings. These discussions took days of involvement from your top people, and the changes suggested had a minimal effect on the big picture. Obvious changes helped to cut costs—they really should have automated the billing department 2 years ago. There is not much motivation to keep meetings interesting. The word in the local union hall is that the TQM process is a synonym for downsizing. Since the payment process was automated, frontline workers lost jobs and there has been little change in the numbers of rejected product. The new equipment keeps breaking down, and your supervisor suspects a new form of industrial sabotage.

Diagnosis of this problem was so tough you hired a management consultant. He said there is a disconnect between the organization's systems and desired behavioral change. Whenever values and systems are not synchronized, stress occurs that evolves into negative attitudes, resentment, and a poor organizational climate. In time, the culture of the organization blocks all change because of these fears and perceptions. New initiatives are difficult to sell. The consultant says converting a negative climate to a positive one is an arduous task. In the past, external threats, such as foreign competition, were used as the catalyst for change. Yet once the organization realized that the threat was not one of life or death, the old systems took over and the negative climate returned.

TQM and Six Sigma, as catalysts for change, suffered from the lack of alignment strategies. New, quick and easy solutions—the low-hanging fruit—were adopted, and a stall occurred that could take years to remedy. What is now needed is a new form of catalyst for change that galvanizes external stimuli (i.e., customer satisfaction) with internal forces (i.e., order fulfillment). When these forces are combined, change that is substantive can be achieved.

The framework for change needs to be one in which learning and continuous improvement elements of Organizational Alignment are accepted. It is one where continuous monitoring is facilitated by a relevant and responsive information system. Creativity needed for breakthrough change is reinforced when the invisible shackles of old values embedded in outdated software systems are surfaced and eradicated. When managers are rewarded for adapting to change requirements and for achieving stretch goals, organizations will be able to accomplish breakthrough goals and become World Class competitors.

The purpose of this book is to define and outline the term *Organizational Alignment* as a means for designing a catalyst for change that is more than superficial. By defining goals that are integrative and "macro" in nature, the kinds of solutions that emerge will yield tangible change and transform systems. When theory and experience come together, great things can happen. The book contains a conceptual overview, along with specific examples of organizations having applied these strategies and a review of the activities needed to apply these Organizational Alignment concepts and strategies in your organization.

The driver for the next wave of customer-centric innovation requires organization-wide alignment and integration of processes, applications, and systems on an unprecedented scale. We call this organization-wide integration "Organizational Alignment," the organizational foundation that can support business in the global economy. It is forcing organizations to ask three key questions:

1. How will alignment change our customer priorities?
2. How can we construct a business design to meet these new customer priorities?
3. What technology investments must we make to survive, let alone thrive?

Note: Throughout this book we will be using the word "business" in a general way to refer to how any organization operates to accomplish its assigned mission and goals. It applies to all organizations—profit, non-profit, and government organizations.

Some of the most damaging and pervasive misconceptions about managing alignment transitions are that you somehow have to wipe out everything that's not working properly and start anew. Put into simpler words: if you're shifting the strategic direction of your organization, must you hire a ton of new expertise from the outside to get you where you are going? In our experience, most organizations that have been successful at managing alignment acceleration do just the opposite. Rather than seek what they need from outside or totally reinvent how they do their work, they take a much more practical and long-lasting approach to getting their organizations aligned by doing it the old-fashioned way: they work with what they've got. This is not to say that these organizations never turn to bringing in new people with fresh ideas or cultivate new ways to create the changes that they need, but rather than search for the silver bullet or magic pill, they work hard to create alignment and bring in outside expertise only as needed.

Aligned organizations stop doing what comes naturally and redefine their future. The philosophy is that every day is a new beginning, with new opportunities and challenges, and tradition is redefined each day and is not an obstacle as it is in most organizations today. Aligned organizations influence and control their environment and are truly seamless learning organizations that prosper and grow while the competition decays. They set their sights high and are masters of boundary-less, fresh out-of-box thinking. They recognize and practice the difference between reengineering and rearranging the pieces. And they benefit from change and refuse to become victims. Instead, they become the masters of transition.

> When your company talks about growth, do the line managers and staff actually know what this means? Is there alignment of their actions?
>
> **—Marc Johnstone, founder, Shirlaws Global**

Organizational Alignment combines today's most powerful improvement methodologies into a single, coordinated initiative, eliminating the guesswork about when and how to apply these tools and eliminating months from the time it typically takes to implement them. Whether it is the design of a new product, service, or process, the enhancement of existing operations, or the development and deployment of new technology, alignment can provide the structure and approach to create a competitive advantage through a focus on customer value and maximizing efficiencies.

This book was written for the organization that has a Strategic Plan and wants to ensure that the total organization is aligned behind it and that it has the very best chance of being successful. We will discuss how to do a Strategic Plan but not in detail. For more detailed information, we suggest reading the book *Organizational Master Plan*, by H. J. Harrington and F. Voehl (CRC/Productivity Press, 2012).

> If your water pipes are out of alignment, you lose a lot of water pressure and water itself. The same is true of an organization that is not aligned with its Strategic Plan.
>
> **—HJH**

Acknowledgment

We want to acknowledge Candy Rogers, who converted and edited endless hours of dictation and misspelled words into the finished product. We couldn't have done it without her dedication, help, and proofreading.

We also would like to acknowledge the many organizations whose senior management and boards of directors we have worked with over the past four decades have helped to form the frameworks, concepts, ideas, and methods outlined in this book. We would also like to thank the many sponsors that helped us drive the research forward by participating in meetings, audio and video conferences, and round-table meetings.

In addition, we would like to thank the American Society for Quality (ASQ) and the International Academy for Quality for their unfailing support for advancing many of these concepts on the national and international levels. Finally, this book is dedicated to all those business leaders who wish to explore this brave new world of organizational alignment, with the help of experienced guides and counselors, in order to discover what works and what doesn't, and to apply the lessons learned to their own organizational and business networks.

—HJH & FWV

About the Author
—H. James Harrington

In Amy Zuckerman's book, *Tech Trending*, Dr. Harrington was referred to as "the quintessential tech trender." The *New York Times* referred to him as having a "knack for synthesis and an open mind about packaging his knowledge and experience in new ways—characteristics that may matter more as prerequisites for new-economy success than technical wizardry."

Well-known business management practices author Tom Peters stated, "I fervently hope that Harrington's readers will not only benefit from the thoroughness of his effort but will also 'smell' the fundamental nature of the challenge for change that he mounts." William Clinton, past president of the United States, appointed Dr. Harrington to serve as an Ambassador of Good Will. It has been said about him, "He writes the books that other consultants use."

Harrington Institute was featured on a half-hour TV program, *Heartbeat of America,* which focuses on outstanding small businesses that make America strong. The host, William Shatner, stated, "You [Dr. Harrington] manage an entrepreneurial company that moves America forward. You are obviously successful."

PRESENT RESPONSIBILITIES

Dr. H. James Harrington now serves as the chief executive officer (CEO) for the Harrington Institute and Harrington Middle East. He also serves as the chairman of the board for a number of businesses.

Dr. Harrington is recognized as one of the world leaders in applying performance improvement methodologies to business processes. He has an excellent record of coming into an organization, working as its CEO or chief operating officer (COO), resulting in a major improvement in its financial and quality performance.

PREVIOUS EXPERIENCE

In February 2002 Dr. Harrington retired as the COO of Systemcorp A.L.G., the leading supplier of knowledge management and project management software solutions, when Systemcorp was purchased by IBM. Prior to this, he served as a principal and one of the leaders in the Process Innovation Group at Ernst & Young; he retired from Ernst & Young when it was purchased by Cap Gemini. Dr. Harrington joined Ernst & Young when Ernst & Young purchased Harrington, Hurd & Rieker, a consulting firm that Dr. Harrington started. Before that, Dr. Harrington was with IBM for over 40 years as a senior engineer and project manager. Dr. Harrington is past chairman and past president of the prestigious International Academy for Quality and of the American Society for Quality Control. He is also an active member of the Global Knowledge Economics Council.

CREDENTIALS

H. James Harrington was elected to the honorary level of the International Academy for Quality, which is the highest level of recognition in the quality profession.

H. James Harrington is a government-registered Quality Engineer, a Certified Quality and Reliability Engineer by the American Society for Quality Control, and a Permanent Certified Professional Manager by the Institute of Certified Professional Managers. He is a Certified Master Six Sigma Black Belt and received the title of Six Sigma Grand Master. H. James Harrington has an MBA and PhD in engineering management and a BS in electrical engineering.

H. James Harrington's contributions to performance improvement around the world have brought him many honors. He was appointed the honorary advisor to the China Quality Control Association, and was elected to the Singapore Productivity Hall of Fame in 1990. He has been named lifetime honorary president of the Asia-Pacific Quality Control Organization and honorary director of the Associación Chilena de Control de Calidad. In 2006 Dr. Harrington accepted the honorary chairman position of Quality Technology Park of Iran.

H. James Harrington has been elected a Fellow of the British Quality Control Organization and the American Society for Quality Control. In 2008 he was elected to be an Honorary Fellow of the Iran Quality Association and Azerbaijan Quality Association. He was also elected an honorary member of the quality societies in Taiwan, Argentina, Brazil, Colombia, and Singapore. He has presented hundreds of papers on performance improvement and organizational management structure at the local, state, national, and international levels.

RECOGNITION

- The Harrington/Ishikawa Medal, presented yearly by the Asian Pacific Quality Organization, was named after H. James Harrington to recognize his many contributions to the region.
- The Harrington/Neron Medal was named after H. James Harrington in 1997 for his many contributions to the quality movement in Canada.
- The Harrington Best TQM Thesis Award was established in 2004 and named after H. James Harrington by the European Universities Network and e-TQM College.
- Harrington Chair in Performance Excellence was established in 2005 at the Sudan University.
- Harrington Excellence Medal was established in 2007 to recognize an individual who uses the quality tools in a superior manner.

H. James Harrington has received many awards, among them the Benjamin L. Lubelsky Award, the John Delbert Award, the Administrative Applications Division Silver Anniversary Award, and the Inspection Division Gold Medal Award. In 1996, he received the ASQC's Lancaster Award in recognition of his international activities. In 2001 he received the Magnolia Award in recognition for the many contributions he has made in improving quality in China. In 2002 H. James Harrington was selected by the European Literati Club to receive a lifetime achievement award at the

Literati Award for Excellence ceremony in London. The award was given to honor his excellent literature contributions to the advancement of quality and organizational performance. Also, in 2002 H. James Harrington was awarded the International Academy of Quality President's Award in recognition for outstanding global leadership in quality and competitiveness, and contributions to IAQ as Nominations Committee chair, vice president, and chairman. In 2003 H. James Harrington received the Edwards Medal from the American Society for Quality (ASQ). The Edwards Medal is presented to the individual who has demonstrated the most outstanding leadership in the application of modern quality control methods, especially through the organization and administration of such work. In 2004 he received the Distinguished Service Award, which is ASQ's highest award for service granted by the ASQ. In 2008 Dr. Harrington was awarded the Sheikh Khalifa Excellence Award (United Arab Emirates) in recognition of his superior performance as an original Quality and Excellence Guru who helped shape modern quality thinking. In 2009 Harrington was selected as the Professional of the Year (2009) by *Quality Magazine*. Also in 2009 he received the Hamdan Bin Mohammed e-University Medal.

CONTACT INFORMATION

Dr. Harrington is a prolific author, publishing hundreds of technical reports and magazine articles. For the past 8 years he has published a monthly column in *Quality Digest Magazine* and is syndicated in five other publications. He has authored 35 books and 10 software packages.

You may contact Dr. Harrington at

16080 Camino del Cerro, Los Gatos, California, 95032
Phone: (408) 358-2476
E-mail: hjh@harrington-institute.com

About the Author
—Frank Voehl

PRESENT RESPONSIBILITIES

Frank Voehl is the Director of Process Innovation at Nova Southeastern University, where he is also a Senior Professor of Executive Education. He also serves as the chairman and president of Strategy Associates, Inc. and a senior consultant and chancellor for the Harrington Institute. He also serves as the chairman of the board for a number of businesses and as a master black belt instructor and technology advisor at the University of Central Florida in Orlando. He is recognized as one of the world leaders in applying quality measurement and Lean Six Sigma methodologies to business processes.

PREVIOUS EXPERIENCE

Frank Voehl has extensive knowledge of NRC, FDA, GMP, & NASA quality system requirements. He is an expert in ISO-9000, QS-9000, ISO-14000, and Six Sigma Quality System Standards and processes. He has degrees from St. John's University and advanced studies at NYU, as well as a Doctor of Divinity degree. Since 1986, he has been responsible for overseeing the implementation of Quality Management systems with organizations in such diverse industries as telecommunications and utilities; federal, state, and local government agencies; public administration and safety; pharmaceuticals; insurance and banking; manufacturing; and institutes of higher learning. In 2002 he joined the Harrington Group as the chief operating officer (COO) and executive vice president. He has held executive management positions with Florida Power and Light and FPL Group, where he was the founding general manager and COO of QualTec Quality Services for 7 years. He has written and published or co-published over 25 books and hundreds of technical papers on business management, quality improvement, logistics, and teambuilding, and has received numerous awards for community leadership, service to third-world countries, and student mentoring.

CREDENTIALS

The Bahamas National Quality Award was developed in 1991 by Voehl to recognize the many contributions of companies in the Caribbean region, and he is an honorary member of its board of judges. In 1980 the City of Yonkers, New York, declared March 7 as Frank Voehl Day, honoring him for his many contributions on behalf of the youth in the city where he lived and performed volunteer work. In 1985 he was named Father of the Year in Broward County, Florida. He also serves as president of the Broward County St. Vincent de Paul Society, whose mission is to serve the poor and needy.

Frank's contributions to quality improvement around the world have brought him many honors and awards, including ASQ's Distinguished Service Medal, the Caribbean Center for Excellence Founders Award, the Community Quality Distinguished Service Award, the Czech Republic Outstanding Service Award on behalf of its business community leaders, FPL's Pioneer Lead Facilitator Award, the Florida SFMA Partners in Productivity Award, and many others. He was appointed the honorary advisor to the Bahamas Quality Control Association, and he was elected to the Eastern Europe Quality Hall of Fame. He has been named honorary director of the Associación Venezuela de Control de Calidad by Banco Consolidado.

Other Books by H. James Harrington and/or Frank Voehl

H. James Harrington and Frank Voehl have published hundreds of technical reports and magazine articles. They have authored over 45 books, most of which are listed here:

- *The Improvement Process*; 1987—one of 1987's best-selling business books
- *Poor-Quality Cost*; 1987
- *Excellence—The IBM Way*; 1988
- *The Quality/Profit Connection*; 1988
- *Business Process Improvement*; 1991—the first book on process redesign
- *The Mouse Story*; 1991
- *Of Tails and Teams*; 1994
- *Total Improvement Management*; 1995
- *High Performance Benchmarking*; 1996
- *The Complete Benchmarking Workbook*; 1996
- *ISO 9000 and Beyond*; 1996
- *The Business Process Improvement Workbook*; 1997
- *The Creativity Toolkit—Provoking Creativity in Individuals and Organizations*; 1998
- *Statistical Analysis Simplified—The Easy-to-Understand Guide to SPC and Data Analysis*; 1998
- *Area Activity Analysis—Aligning Work Activities and Measurements to Enhance Business Performance*; 1998
- *Reliability Simplified—Going beyond Quality to Keep Customers for Life*; 1999
- *ISO 14000 Implementation—Upgrading Your EMS Effectively*; 1999
- *Performance Improvement Methods—Fighting the War on Waste*; 1999
- *Simulation Modeling Methods—An Interactive Guide to Results-Based Decision Making*; 2000
- *Project Change Management—Applying Change Management to Improvement Projects*; 2000
- *E-Business Project Manager*; 2002

- *Process Management Excellence—The Art of Excelling in Process Management*; 2005
- *Project Management Excellence—The Art of Excelling in Project Management*; 2005
- *Change Management Excellence—The Art of Excelling in Change Management*; 2005
- *Knowledge Management Excellence—The Art of Excelling in Knowledge Management*; 2005
- *Resource Management Excellence—The Art of Excelling in Resource Management*; 2005
- *Six Sigma Statistics Simplified*; 2006
- *Improving Healthcare Quality and Cost with Six Sigma*; 2006
- *Six Sigma Green Belt Workbook*; 2007
- *Six Sigma Yellow Belt Workbook*; 2007
- *Making Teams Hum*; 2007
- *Advanced Performance Improvement Approaches: Waging the War on Waste II*; 2007
- *ISO 9000: An Implementation Guide for Small to Mid-Sized Businesses* (with Peter Jackson and David Ashton, St. Lucie Press, 1994)
- *Deming: The Way We Knew Him* (CRC Press, 1995)
- *Handbook for TQM Implementation* (St. Lucie Press, 1994)
- *Teambuilding: A Structured Learning Approach* (with Peter Mears, CRC Press, 1995)
- *The Executive Guide to Implementing Quality Management Systems* (with Peter Mears, CRC Press, 1995)
- *Macrologistics Management: A Catalyst for Organizational Change* (with Martin Stein, CRC Press, 1997)
- *Total Quality in Information Systems and Technology* (with Jack Woodall and Deborah K. Rebuck, CRC Press, 1996)
- *Problem Solving for Results* (with Bill Roth and James Ryder, St. Lucie Press/CRC Press, 1996)

1

Overview

Each time there is a major change to the Strategic Plan, the management team should consider doing an Organization Alignment exercise.

—HJH

WHY ORGANIZATIONAL ALIGNMENT?

In an effective organization there is congruence between purpose, strategy, processes, structure, culture and people. It is the challenge [of] the leaders to orchestrate this alignment and to still promote innovation and change.

—**David J. MacCoy**

Definition: *Organizational Alignment* is the methodology that brings the organization's structure, processes, networks, people, and reward system in harmony with the Strategic Business Plan and the Strategic Improvement Plan. It is a strategic management process which harnesses the energy of all employees to contribute to the accomplishment of corporate objectives.

Organizational Alignment is the linking of strategic, cultural, processes, people, management, systems, and rewards to accomplish the best results. It occurs when strategic goals and cultural values are mutually supportive and where each part of the organization is linked and compatible with each other. In other words, an effective organization develops a tight fit between purpose, strategy, processes, structure, culture, and people, and it is the challenge for the leaders to orchestrate this alignment and to still promote innovation and change. Many organizations have failed at process redesign, total quality management (TQM), Six Sigma, and other continuous improvement efforts because they failed to recognize the impact and importance of behavioral alignment on sustained performance. They worked very hard at copying each other's improvement programs, but they have not really changed the way people interact and view their workplace on a day-to-day basis.

The organization is made up of a complex sensory system similar to our body's nervous system: one action in one part of the system can cause a major reaction in other systems within the organization—just as putting your hand on a hot surface causes your arm to jerk back, your brain to register pain, your heart to pump blood faster, and your vocal chords to let out a scream followed by a few choice words.

THE ORGANIZATIONAL ALIGNMENT FUNCTIONAL MODEL

The dynamic system that makes up your organization functions at its peak performance when these systems are designed to work in harmony. To accomplish this, the organization's culture and strategies must not only be in harmony, but they must also be complementary so that they produce the optimum results. Picture the cultural makeup of the organization and the strategies as two different paths running in parallel with each other, as is shown in the Organizational Alignment Functional Model (see Figure 1.1).

The strategic path defines what the organization wants to accomplish down to the activities and tasks levels that the individual will need to perform. The cultural path defines how things should be done, modified by the day-to-day behavior of the management team and the experience of the employees. At the pinnacle is the organization's mission and vision.

> Definition: *Mission* is the stated reason for the existence of the organization. It is usually prepared by the chief executive officer (CEO) and seldom changes, normally only when the organization decides to pursue a completely new market.
> Definition: *Vision* is a documented or mental description or picture of a desired future state of an organization, process, team, key business drivers (KBD), or activity.

FIGURE 1.1
The Organizational Alignment Functional Model.

Definition: A *Long-Term Vision Statement* for the organization is usually prepared by top management and defines the desired future state
of an organization 10 to 25 years in the future from the date that the
statement was created.

Definition: A *Short-Term Vision Statement* for the organization is usually prepared by top management and defines the desired future state
of an organization 5 or 10 years in the future from the date that the
statement was created.

Definition: *Value Statements* are documented directives that set
behavioral patterns for all employees. They are deeply engrained
operating rules or guiding principles of an organization that should
not be compromised. Value statements are sometimes called operating principles, guiding principles, basic beliefs, or operating
rules.

Definition: *Outcomes* are the measured results that the organization
realized as a result of the action(s) taken.

The mission statement is essential to linking the organization with its
vision of the future. Some organizations call this their "purpose statement"
or the central reason why they are in business. A good mission statement
defines the type of product or service the organization will provide. It is
used to determine if an opportunity is in or outside of the scope of the
organization. It can be worded as a "to be" or "to do" statement. Either
type of mission statement can be effective, so the academics or consultants' arguments over which one is correct are unnecessary.

What's the difference between a *vision* and a *mission* for an organization or with the *purpose* of the organization? Many organizations go through an agonizing process of trying to determine the distinction based on vague definitions offered by consultants and academics. We have spent hours of valuable time discussing which comes first: the mission or vision statement. We believe that you define the type of business you are in (mission statement) and then define how you want the organization to be viewed in the future (vision statement), but doing it the other way around also works. Could it be *vision and mission*? Instead of trying to split the "definitional hairs," we have found that a useful perspective can be gained by utilizing the approach of linking your organization's internal efforts with the external world in which you compete and serve customers.

In terms of Organizational Alignment, having a compelling vision of the future of the market and the industry is one of the key implementation challenges and a vital ingredient in an organizational plan. By "vision" we mean a view of what the organization will be like 5 to 20 years from now. It could be as simple as "an affordable, easy-to-use personal computer on everyone's desk" or "news available immediately from anywhere in the world." The winners tend to be able to express an energizing picture of the future in terms of market presence and customer benefits and have enough reality to it to make it aggressively believable. The losers tend to lack any vision and exist, from day to day, reacting to the market and the leads of other competitors.

For example, escalating deregulation of the electric utility industry was opening up local markets to all sorts of new competition from other former regional monopolies and newly independent electric power brokers in terms of market presence and customer benefits. The Southern Company's executives needed to reposition their huge organization for much greater competition in their traditionally regulated industry. In spite of the fact that Southern Company executives had plenty of resources to support a major transformation realignment effort, representatives from all parts of the company needed to expand their vision in order to be dramatically exposed to the reality of totally new competitors.

Winners make their mission and vision statements short, clear, and compelling, while losers will have mission and vision statements focused on shareholder value or some other noncustomer, noncompetitive emphasis.

Key Implementation Challenges

The leadership of Southern Pacific needed to impress upon employees the need for change beyond incremental extensions and refinements of the current process. Substantial effort should be directed at the unionized employees so that they understand the need for fundamental change in order to be aligned with the corporate transformation that is required.

Another example is the National Semiconductor Company. Once one of the excellent companies in the classic book *In Search of Excellence,* at one point in time National had only a few days of cash, its banking facility had been revoked, and it was about to write off $150 million for the second year in a row. As a result, resource deprived was this company that its leadership was forced to sell one of the organization's most modern plants in order to generate cash to jump-start its corporate transformation process.

There was little doubt that after having incurred significant losses in four of the five previous years, the managers and employees of National Semiconductor Company (once labeled the "animals of Silicon Valley" for their prowess in pursuing a low-cost, high-volume, chip-making strategy) were ready for revitalization through corporate realignment. Everyone was eager for fundamental change, but how to do it, and with what? The primary challenges posed by the company's initial change condition, therefore, were those of resource creation and focusing and aligning corporate energy among executives and employees for corporate "culture change" transformation.* The results of these turnarounds using alignment tools and techniques will be discussed in later chapters.

Winners embody a strong leadership emphasis in their planning, as reported in a study of almost 300 organizations in Ernst & Young's American Competitiveness Study. This study was done in conjunction with the editors of *Electronic Business.* It was written by John H. Sheridan and published in *Industry Week,* May 21, 1990. This emphasis on being a leader is essential to both the development of strategy and to motivating

---

* The phrase "corporate culture" or "organizational culture" does not appear in James G. March's timeless classic *Handbook of Organizations* (Chicago: Rand McNally, 1965) or in Peter Drucker's *Management: Tasks, Responsibilities, Practices* (New York: Harper & Row, 1974). Even the word "culture" is hard to find in Organizational Alignment literature, although some of the ideas we now associate with this concept were developed under the heading of "norms and values [in small groups]." A classic example is George Homans's *The Human Group* (New York: Harcourt, Brace & World, 1950). The first scholarly work to focus on organizational culture appeared in 1979; see Andrew Pettigrew's "On Organizational Culture," *Administrative Science Quarterly* 24 (1979): 570–581. Some of these "older" scholarly works can be found in the "Golden Oldie" Additional Resources section at the end of each chapter, as appropriate.

the organization's people, customers, and suppliers by focusing on being a winner and not a follower. Staying the course involves the alignment of key critical systems and critical cultural philosophies. This alignment helps produce the desired behaviors in the organization needed to produce the organization's outcomes to achieve the mission and vision.

> An organization that is not aligned is like a train running on rails that are not aligned.

> —HJH

Gauging Effectiveness

Execution or staying the course is often a very difficult task in that along the way we often encounter many potential obstacles to failure. Leadership in an aligned organization ensures that (1) all the people understand what is required of them, (2) their actions support it, and (3) everyone is held accountable for his and her actions.

A survey is a way for an organization to track and project how past, current, and potential organizational milestones have affected or will affect the organization, with positive or negative energy, each time a major milestone is reached.

Organizational milestones can be a product or service introduction, a corporate reorganization or change in leadership, a merger or takeover, or a technological breakthrough. What follows is the Reality Check Survey. A survey by *Inc.* magazine showed that most managers have a long way to go to accomplish these goals. In addition, 8 out of 10 employees said that they are not held accountable for their actions.*

ALIGNMENT TOOL #1: REALITY CHECK SURVEY

The following is a survey that we have found useful.

Instructions: Please read each statement and indicate the extent to which it describes the reality in your organization as a whole. Your responses should reflect what you have experienced as well as what you have generally observed.

---

* The *Inc.* magazine survey, first conducted in March 1993, has been repeated every 5 years or so in informal follow-ups. See Inc. 15, no. 3 (1993): 34.

Rate each statement using a 10-point scale. The left side of the scale indicates that you *totally disagree* and the right side that you *strongly agree*. If you do not know, indicate so next to the question.

Please take the time to respond to the open-ended questions at the survey end, as your responses are critical in improving the readiness for change. Be honest in your responses as there is no right or wrong answer. Your responses will remain completely confidential.

1. My organization has a clear written mission, vision, and values statement(s).
2. This statement(s) is supported by management actions.
3. Our organization must change the way it works.
4. All departments, divisions, and processes have measurable goals.
5. The leaders of our organization seem committed to an immediate change.
6. Every employee understands what is expected in terms of performance.
7. To get people motivated when undertaking a change, leadership focuses on more than the logical business case.
8. All employees are held accountable for daily performance.
9. Employees believe that we cannot stay ahead by continuing to work exactly as we do today.
10. My boss understands my job well enough to guide me in changing the way I work.

Totals for Reality Check: Add up the scores for each of the 10 statements: _____

Note: Grand Total of 50 or less indicates very serious problems and 80 to 100 indicates no serious issues, while 50 to 80 indicates need for improvement.

Now Ask Yourself These Questions*:

- What am I doing to create and understand this situation?
- Are my emotions getting in the way, whether helping me or hurting me?

---

* Adapted from Robert F. Sarmiento, *Reality Check: Twenty Questions to Screw Your Head on Straight* (Houston, TX: Bunker Hill Press, 1993).

- What am I telling myself in my heart-of-hearts?
- What are the real facts, and am I exaggerating or distorting them?
- Are there other explanations that are lying dormant?
- How likely are my worries, and whose problem is this anyway?
- What is the worst that can happen?
- Am I taking this too seriously? Too personally?
- Am I doubting myself or others?
- Am I unrealistically demanding success? Approval? Control? Perfection? Certainty? Comfort? Fairness? My way?
- Do I really need this or only want it?
- Can I stand it, or am I babying myself?
- How can I begin to think more realistically?
- Am I stewing rather than doing?
- What are my options and what, if anything, can I do?

Reality Check for Worriers
- Am I playing "what if"?
- How probable are my worries?
- Am I worrying enough?
- Will worrying do any good?
- Do I equate worry with caring?
- What, if anything, can I do?

Reality Check for Procrastinators
- What emotions do I have about what I am putting off?
- What are the advantages of doing it now?
- What are the disadvantages of procrastinating?
- Am I making excuses?
- Am I putting myself down for procrastinating?
- Will it kill me to do it for 5 minutes?
- Can I break this task down?
- What consequences can I give myself?
- Am I doing the worst first?
- Am I babying myself?

HARNESSING THE ENERGY OF ALIGNMENT: WITH ALBERT ("BUTCH") EINSTEIN AND JACK ("SUNDANCE") WELCH

Butch Cassidy and the Sundance Kid is a 1969 high-energy film about two western bank and train robbers who flee to Bolivia when the law gets too close. During a gunfight with the Bolivian police, Butch and Sundance run low on ammunition.

Butch Cassidy: We're going to run out unless we can get to that mule and get some more.
Sundance Kid: I'll go.
Butch Cassidy: This is no time for bravery. I'll let ya!

When we think of management gurus, whereas the name Jack Welch resonates loudly, the name Albert Einstein does not. But the revered theoretical physicist can teach us volumes about real bravery and how to lead an organization into the 21st century. Like Einstein, Jack Welch and other leaders of the future understood that they must know the elements they are working with, have high aspirations and ideals, and understand the dynamics of movement and how they create energy.

To prevent their vision from becoming a mere slogan, organizations need to show employees how they can translate abstraction into the specific energy and meaningful language they need to get through the period of Organizational Alignment. Energy and meaningful language, not money or ideas, are the scarce resources of behavior-driven change. Consider energy (or, if you prefer, effort). It takes extra energy for an individual to work through the sources of reluctance. Neither understanding nor desire nor planning nor action can come without effort, and people rarely progress without setbacks. Instead, they must work hard throughout the period of change until they have integrated new behaviors into daily routines.

In a real sense, people confronting behavior and skill change at work have two jobs: a "from" job and a "to" job. People know the "from" job well—it is completing the variety of tasks and goals assigned to them.

The second job is learning how to work differently.* Ultimately, the individuals must integrate new behaviors and skills. As previously mentioned, Organizational Alignment is a strategic management process that harnesses the energy of all employees to contribute to the accomplishment of corporate objectives. This process works as a balance point and interface between employee performance, organizational performance, and customer delight. The measurement system Balanced Scorecard (BSC) is one of the tools deployed by some organizations to achieve this alignment.

Policy Deployment (Hoshin Kanri) and Quality Function Deployment (QFD) are other approaches to align employees and processes with customer needs and expectations. Through Policy Deployment, the organization launches certain initiatives every year to achieve their corporate objectives. These initiatives are translated into a set of measurable tasks that are assigned to teams as well as individual employees.

The Concept of Organizational Alignment

The concept of Organizational Alignment considers the extent to which organizational strategy, structure, and culture create an environment that facilitates achievement of organizational objectives and development of high-performance work organizations. Well-aligned organizations have systematic agreement among goals, tactics, reward systems, and culture. Imagine that you are the CEO of that Fortune 500 company previously discussed. You have just left a board meeting where the members expressed their dissatisfaction with progress achieved last year. In your head you know that there is a rational explanation for the failure of the organization to change; in your heart you feel that there are insurmountable obstacles.

The framework for change needs to be one where learning and continuous improvement elements are accepted. It is one where continuous monitoring is facilitated by a relevant and responsive information system. Creativity needed for breakthrough change is reinforced when the

---

* Ray Statta, "Organizational Learning—The Key to Management Innovation," *Sloan Management Review* (Spring 1989): 63–74; also see James Higgins's seminal work on alignment, *Innovate or Evaporate* (New Management, Winter Park, FL, 1995) in which he describes in detail the four types of Organizational Alignment required for innovation, along with a description of the culture change required to foster innovation among all employees. See also Paul Plsek, *Creativity, Innovation, and Quality* (ASQC Quality Press, Milwaukee, WI, 1997).

invisible shackles of old values embedded in outdated software systems are surfaced and eradicated. When managers are rewarded for adapting to change requirements and for achieving stretch goals, organizations will be able to accomplish breakthrough goals and become world-class competitors.

- Energy is an organization's capacity for action and accomplishment. According to Bill Plamondon, past CEO of Budget Rent-a-Car,* "it propels the organization forward, maintains its balance, and keeps it focused during downturns, transitions, and crises." He asserts that it is not enough for leaders simply to possess energy; the leader's job is to help others in the organization generate their own energy and pass it on. If we apply Einstein's famous statement of the mass-energy relationship, $E = mc^2$ (energy equals mass times the speed of light squared), to an organization, we could say that energy (E) is created by leaders who inspire the members of their organizations (m) to anticipate and respond at high velocity (c^2).

- Einstein's special theory of relativity applies to business physics. The basic theory reflects a positive correlation between mass and energy. Applied to organizations, Einstein's laws tend to behave in the opposite manner. Large organizations can get bogged down in bureaucracy and rules that frustrate and demotivate their members. Processes they may have instituted to help employees make responsible decisions can become ends unto themselves—until the way things are done supersedes getting them done. Leaders of these organizations often believe that if they build enough procedural protection, no mistakes will be made and the right decisions will be ensured. But they accomplish the opposite.

- Decisions, when they are made at all, are made by default because no one is accountable, because no individual or group takes responsibility. The issue of not taking responsibility becomes institutionalized in layers of rules, forms, and unproductive meetings. Bureaucracy and a fear of making mistakes indicate that the organization's management

---

* William N. Plamondon is the past president and CEO of Budget Rent-a-Car Corporation. He serves on the board of directors of the American Car Rental Association; Northern Central College in Naperville, Illinois; and the Florida-based "Give Kids the World," for which he also actively volunteers. Plamondon is active in the International Franchise Association, the White House Council on Travel and Tourism Issues Task Force, and the World Travel and Tourism Council. He encouraged all employees to "aim high, play it straight, and make it fun."

structure and processes have ceased to be effective, leaving no alternative but to realign them to support an environment that breeds energy.

- The energy-breeding environment. But what does such an environment look like? In an aligned organization world, it is one where four fundamental building blocks occur:
  - The organization needs to stay open to environmental information from customers, employees, competitors, and the marketplace.
  - The leadership team is aware of its strengths and weaknesses compared to the strengths and weaknesses of the competition and plays within them.
  - Employees have a well-developed sense of purpose beyond just making money, which is guided by a core ideology as well as compelling and challenging performance goals.
  - There are many leaders. Authority and accountability are decentralized, so that the organization becomes a collection of small, interchangeable units working toward a common goal.*
- The organization stays open to environmental information. An organization is more than the sum of its people, products, and capital; it contains a unique type of DNA and is organically aligned. It has a life of its own and to grow and prosper, it needs to remain open to the environment and the signals of the market. By doing so, the organization and its systems and culture will stay resilient and organized around stakeholder requirements, while maintaining open communication with the entire stakeholder community: customers, suppliers, and employees. By cultivating this kind of participation, leaders relinquish some of what is traditionally thought of as control but in the end create a long-lasting aligned organization, because new energy is created when employees are given authority and responsibility.
- Know the leadership team's strengths and weaknesses and play within them. In addition to listening to customers and suppliers and staying abreast of current industry and market trends, an aligned organization possesses a sense of history. If leaders don't know history, then they won't be able to begin to visualize and understand the

---

* These four building blocks are loosely based upon Bill Plamondon, Ch. 28, Energy and Leadership, in *The Leader of the Future,* Frances Hesselbein, Marshall Goldsmith and Richard Beckhard, eds., Peter Drucker Foundation, NY, 1996. This book is a classic management sampler, comprising essays from academics and corporate CEOs about leadership, or specifically the kind of leadership and management that will be required for successful results of organizations in the future.

future. A sense of history gives leaders a greater perspective on their organization's and leadership strengths and weaknesses, as leaders begin to see the inevitable cycles of business and recognize the things they can do to prolong the upswing of the cycles, while discovering ways to make the most of the downtimes without turning the organization against itself or burning employees out. Leaders who are resilient bounce back quickly and learn to recharge the energy of their employees, rather than consume it, in order to keep the organization moving ahead.*

- Employees develop a sense of purpose and a higher goal. Aligned organizations understand that they are held together by shared visions, values, beliefs, and obligations, which is what enables them to rise above cyclical hardships and gives it its integrity and capacity to endure. Once employees embrace and understand the organization's core values, the most effective way to unify the organization and create momentum is to commit to a clear and compelling goal that anticipates market needs and is aligned with what the organization stands for, a goal that stimulates energy because it is tangible, resonant, highly focused, and compelling. It is the leaders' responsibility to communicate this goal in a clear and compelling way that inspires the organization to move to new heights and at faster speeds than it would ordinarily attain on its own.†

- Create many leaders by decentralizing authority and accountability. The key element of an aligned energy-breeding organizational environment is the presence of many leaders from bottom to top. Leadership can become a heavy burden and it needs to be shared. As more people become leaders, the organization will be able to grow, respond, and move faster and faster, thus creating more energy, as Einstein stated: $E = mc^2$. The leader of the future will need to guide and mentor the organization's people through these changes. And whether that leader is in a board room or the back room, the skill sets he or she will be called to draw upon will often intersect with Einstein's qualities of wisdom, vision, and the ability to create energy. So follow Jack Welch's business model: hire and reward the right people; instill in them the core values of the organization; give

---

* Ibid.
† Ibid.

them a clear target goal to shoot for, along with accountability and responsibility; and then get the heck out of their way.

Lessons from "Butch" Einstein

Among the many lessons we can learn from Einstein's teachings is that energy should be directed toward the good of the organization and its people, and its leaders can find inspiration in his words:

> Each of us comes for a short visit, not knowing why yet sometimes seeming to divine a purpose. From the standpoint of daily life, however, there is one thing we do know: that man is here for the sake of other men—above all for those upon whose smile and well-being our own happiness depends, and also for the countless unknown souls with whose fate we are connected by a bond of sympathy. Many times a day I realize how much my own outer and inner life is built upon the labors of my fellow men, both living and dead, and how earnestly I must exert myself in order to give in return as much as I have received.

—Albert Einstein, *The World as I See It: Ideas and Opinions,* 1954, p. 8.

Lessons from "Sundance" Welch

As the legendary CEO of General Electric, Jack Welch repeatedly stated that he expected his managers to move their company to a new type of flat organization, which meant less vertical hierarchy, which slowed decisions and bureaucratized everything. He wanted to see the scope of an individual manager's efforts to be "boundary-less." This new form of empowered manager would have his success measured in terms of the "speed" by which change is accomplished and results are accelerated. Goals are acceptable only if they stretch the organization's capabilities.

Jack Welch provides for us a wonderful confirmation of the need for an Organizational Alignment strategy. It will take stretch to accomplish significant change. Competitors who prosper will be able to meet stretch goals with speed. They will be able to accelerate the change process and implement stretch goals in a seamless fashion across organizational divisions. In the past, economists and accountants who were largely responsible for design of corporate performance standards focused on tangibles. The accountant wanted performance to be measured in terms of tangible flows such as income and expenses. Success was defined as the condition where there is more income than expense.

Performance measures that incorporate time and place utility, in addition to form utility, will make all the difference in the future.* Accountants are now actively redefining their measures to calculate the "value-added" from process. During the past 20 years or so, there emerged a new basis of accounting called activity-based accounting, which implied that economists would need to define new measures for organizations that succeed by delivering more than form utility. When radical improvements, breakthroughs, in time and place utility occur, it can result in the redefinition of an industry. For example, when Federal Express, perhaps the pioneer of Organizational Alignment, introduced an overnight standard, they redefined the small-package delivery industry. Domino's Pizza delivers the same type of pizza as competitors, but the delivery is to your home and with a standard of performance of 30 minutes, which gave them a temporary competitive advantage until the competitors caught up.

Two Lessons from Butch and Sundance

- Stretch goals for speed will be added to the requirement for delivery of goods and services. The extra value added in terms of time reduction will be so large that new measures will be needed. Achievement of these time and place goals, such as in the new standard for manufacturing—Just-in-Time (JIT)—will require huge new investments. How can that new investment be rationalized without better performance measures? Information systems and other support systems will have to adapt to these new stretch goals. What is needed is a new framework for measuring and tracking the success of the organization in delivering on these types of speed- and location-specific goals.
- A new framework is needed to cut across the organizational functions and needs to also consider the role of suppliers and vendors of materials. Further, because environmental regulations require

---

* P. Rajan Varadarajan and Vasudevan Ramanujam, "The Corporate Performance Conundrum: A Synthesis of Contemporary Views," *Journal of Management Studies,* 27 (September, 1990): 463–483; revisited January 2005; James Higgins, *Innovate or Evaporate* (Winter Park, FL: New Management Publishing, 1995). Higgins also presents an integrated approach, which focuses on process innovation, an alignment driver. In this seminal work, he has focused on seven process innovation areas: marketing, operations, finance, HR management, information systems, and technology, research, and development, and management; Plsek takes a similar approach in his work cited in footnote 4. Also see Masaaki Kotabe, "Corporate Product Policy and Innovative Behavior of European and Japanese Multinationals," *Journal of Marketing*, 54 (April, 1990): 19–33.

organizations to dispose or recycle materials used in the manufacturing process, these elements also are important in designing the acceptable framework. Although leaders have often used logistics in the military to fine-tune the process of delivery that is time and/or location specific, there is a significant delay in the incorporation of military logistics requirements in the civilian sector. For example, in the late 1990s, the military used Organizational Alignment in a broader sense. In the Desert Storm war, General H. Norman Schwarzkopf challenged the assumption that helicopters had an assumed downtime of 50%. According to the Theory of Constraints, capacity to move material to the frontline jeopardized the success of the war. Without more rapid turnaround time on helicopters, the resultant rapid deployment of troops in Kuwait could not be achieved. No one could give Schwarzkopf a reason for the assumption, so he changed the assumption to 20% and made it a Big Hairy Audacious Goal (BHAG): a stretch goal. After the downtime expectation was reduced, new maintenance and parts pre-positioning systems were needed, and modifications were made to repair facilities. The new goal was achieved with much less system pain than had been expected.

Managers given this stretch goal were forced to rethink the supply chain, and repair processes were rescheduled. Much like the tire-changing teams at the Daytona or Indy 500, the new "pit" crews for helicopters used better teamwork to speed up the repair process and the results were extremely successful, not just for the war effort but also as a way to demonstrate to military managers that overall alignment goals, at the "theater" or systems-wide level, could be used as a catalyst for change.

ORGANIZATIONAL ALIGNMENT OUTCOMES

A major outcome and advantage of Organizational Alignment is that the process unleashes new sources of energy and synergy, as previously described. These synergies help promote creativity and help facilitate effective and rapid change, thereby creating additional energy within the organization. The sources of Organizational Alignment renewed energy and synergy are as follows:

1. Suppliers—Often suppliers have information, research, and new technology that can help. Currently, many organizations have only one-way dialogues with their suppliers. Opening up a two-way dialogue provides new ideas for change, and the dynamic exchange of information improves productivity. JIT II is an example of a best practice in this area, and several case studies in the following chapters will describe this new approach.

2. In-plants—Also found in the JIT II process, in-plants, on-site, and empowered supplier "resource" managers provide opportunities for concurrent engineering and new ideas on production efficiency. They also serve as troubleshooters and can help prevent problems that can stop the production process.

3. Organization rules for innovation—Once the strategy has been deployed, it is necessary to identify how many rules need to be broken to make it work. Also, are employees and cross-functional managers recommending flexibility in some rules? These may be cases where there are hidden opportunities to leverage rule changes to spark new change. For example, at Xerox the salesperson defines delivery dates. Meeting his expectations was easy because by practice—a kind of organizational rule—he always filled in the desired delivery dates. Managers were able to redefine the process and get the customers' real needs recorded on the order form, forcing the delivery process to meet higher standards of service.

4. Benchmarking using critical success factors—Sending groups of managers out into other organizations helps stimulate new ideas. Particularly if the mix can include diverse organizations, new creative ideas and innovations are likely. Because Organizational Alignment strategies involve considerable new investments in information system change, benchmarking is desirable.

5. Performance measures—Organizational Alignment strategies can only be effective if there is considerable effort spent on the design of new performance measures. The new order fulfillment process when it is benchmarked will have to have specific quantitative measures. These measures force considerable discussion about goals and objectives. This unleashes synergy across organizational lines.

6. Process transformation—The use of a Quality Management process, such as Lean Six Sigma, is one way to start an organizational transformation process and is described in more detail in

the section on integration. This motivation to excel in quality has attracted many supporters because it was successfully used by Japanese organizations to achieve global dominance in industries, such as electronics and the automobile industry. The desire to attain the Malcolm Baldrige Award is one way the federal government hoped to stimulate corporate competitiveness and promote new investment.

CHALLENGES TO IMPLEMENTATION

Corporate managers looked at the point scores in the Malcolm Baldrige Award and tried to achieve high point scores without understanding how the process worked completely. Separate units and teams were given very specific goals. Failure to achieve implementation of the program objectives was one of the outcomes of the blind adherence to literal interpretation of the Baldrige processes. Also, the failure to implement the TQM programs of the past decade as a way to transform organizations were in part due to the inability of managers to see the full impact of their actions and the failure to have a meaningful catalyst for alignment and change.

Lack of empowerment results in superficial changes, and the lack of a catalyst reduces the process to a series of staccato actions. Improvement to process with no unifying objective can result in minor change, but substantial differences in the way an organization's process are performed are less likely. Consequently, the impact on the core organization's outputs is not substantive and there is little impact on the customer or market share. Managers and executives, who see little impact from their organizational improvement process other than higher point totals on an arbitrary measure such as the 1000-point system in the Baldrige scoring process, soon lose motivation for further change.

What is needed is the use of an integrated catalyst for change that is lasting and that can permanently influence the way an organization performs its core processes and delivers its core outputs.* There are recent articles about

---

* Almost everyone writing on this topic supports this theory in passing, but few make it the center of their ideas. Schein, Lorsch, and Davis seem to have leaned toward this model the most. One could argue that this whole perspective comes from early work by Lorsch (see Paul Lawrence and Jay Lorsch, *Organization and Environment* [Boston: Harvard Business School Press, 1967]) or even earlier work by Tom Burns and G. M. Stalker (see *The Management of Innovation* [London: Tavistock, 1961]).

managers who fear loss of employment and who are understandably less enthusiastic about pursuing Lean-based change processes. Organizational alignment offers the promise of new challenges, new markets, and new opportunities. When the challenges are achieved, the result is the avoidance of negative consequences such as market share decline. Now the goal is market expansion by opening up new opportunities for sales. For example, rapid delivery provides a larger geographic market area, thus instantly adding new customers.

Imagine that you approach a half a dozen of your employees and ask them what they do, how it fits into the larger scheme of things, and how their work contributes to the overall accomplishment of your corporate mission and strategy. Do they know the answer? Are you satisfied that they have a clear line of sight that allows them to optimize their contribution? Every organization needs strategies, goals, and objectives. They are the organization's compass, showing which way is True North. Plans and tactics are the road maps that describe the route from here to there. But in complex, highly interdependent organizations, it's not enough for the parts of the organization to have and know only their own direction. They must also know what's going on in other parts of that organization, with whom and in what ways they are interdependent, and how they must work together to ensure overall success. At a minimum, your employees must have access to and understand all of this information to answer your question.

Organizational goals, objectives, assessment, and rewards must reflect this interdependence. This is the "demand signal" for organization and employee alignment and deployment of resources. Your internal or external consultants must work with you to plan and implement processes and events internal to your organization that will improve the synergy and productivity of your people by doing the following:

- Providing them with essential information about the direction, strategy, and intended results of your organization
- Clarifying big picture connections, context, and meaning
- Facilitating individual and functional reflection and planning about the plans and goals, what they mean, and how they can best contribute
- Inviting feedback and dialogue about the plans and goals and what it will take to accomplish them
- Identifying and arranging key interdependencies and points of required collaboration

- Providing regular updates about progress toward goals, key metrics, accomplishments, discoveries, etc.
- Ensuring that organization learning occurs

Neither is it enough to focus just inside the boundary of your own work unit, or even your own organization. Today, organizations are comprised of supply chains—large, complex networks of organizations that are in supplier-customer relationships with each other. Organizations are almost always members of multiple supply chains and play a variety of roles: sometimes the customer, sometimes a supplier to other customers. Perhaps an organization that is your supplier in one supply chain is your competitor in a mutual customer's supply chain.*

How likely are employees at organizations with which you partner and on whom you are dependent to know how what they do enables your success? Do they even know that you're counting on them?

As these roles become increasingly blurred, it's essential that you understand how the organizations, who are players in your supply chain, are interdependent and impact the success or failure of others in the network, and how your organization will eventually be impacted. Goal and strategy alignment, sharing critical (and sometimes confidential) information that will contribute to improved relationships and performance, and joint management of the partnership are key to successful and effective customer-supplier engagements.

New Sources of Competitive Advantage

Designing and effectively managing these incredibly complex networks of relationships is an area of enormous and largely untapped opportunity and a source of competitive advantage, because of the following reasons:

- The prevailing paradigm is competition, not collaboration.
- Most organizations are not designed to enable or facilitate collaboration, whether internal or external.

---

* We are grateful to Patty Anton for sharing with us her thinking on the subject of Organizational Alignment strategies and Supply Chain management. Patty Anton is vice president of marketing effectiveness at Quaero Corporation and has over 15 years of experience helping clients improve the quality and profitability of their customer interactions.

- Data and information do not equal meaning.
- Processes and methods (even those that are known) that will bridge the interfaces at the boundaries are not well utilized.

Working This New Aligned Way

Very few organizations are actively engaged in learning how to truly transform supply chain partnerships to enable them to collaborate effectively with partners (never mind competitors) for true, sustained mutual benefit. Figuring out how to do this requires a customized approach in which participants are cognizant of the unique needs, cultures, and business requirements of the involved organizations and also comprehend the nature of large systems and complex organization change. Here are some of the ways that Organizational Alignment can help you design and optimize your supply chain relationships so that collaboration becomes a source of competitive advantage:

- Ensure that your alignment accurately reflects the cultural, values, behavioral, and joint management requirements of the customer-supplier relationship.
- Make an assessment of the short list of candidate organizations that have met the technical requirements and appear to be most capable of delivering on these other requirements.
- Assess relationships with current and prior suppliers of similar services for use in designing and developing new specifications, processes, and relationships.
- Develop processes to map and effectively use assessment findings in making a balanced decision about the best supplier for the contract; participate in the supplier selection process.
- Design and facilitate a joint planning process that identifies, addresses, and develops interdependencies; points of and processes/structures for collaboration; strategy and goal alignment; goals, metrics, and desired outcomes and processes for assessing and evaluating same.
- Redesign both organizations' infrastructures as necessary to support and sustain the partnership.
- Organize intra- and cross-organization team startup meetings (yours, mine, ours).
- Develop, utilize, and evolve structures and processes for jointly managing interdependencies, communication, problems and conflicts, performance appraisal and rewards, evaluation of results, etc.

Focusing the Alignment Strategy

The requirements for using the Organizational Alignment strategy as a catalyst for a successful value-adding process are the three-phase conceptual framework and competent leadership.

- Conceptual Framework—The change process needs to be divided into three separate phases. Some strategies work well in one phase, and others are needed in later phases. The three phases are alignment, mobilization, and integration. For example, JIT II is an excellent tool for building alignment. The use of business rules analysis is a critical part of transitioning from alignment to mobilization. This strategy works well here because it eliminates obstacles in the rules process that inhibit achievement of goals that are buried in obscure locations, such as software code or obsolete procedures and practices. Adaptive engineering works well when there is alignment and mobilization and the organization can focus on a rapidly improving approach to meeting new stretch goals.

- Leadership—Often overlooked, competent leadership is critical. In some small organizations, owners make poor leaders because their economic power stifles challenge and creativity. Developing change processes requires, as a core competency, leaders who can effectively move organizations from one plateau to another. The leadership process requires CEO-level ingredients, such as a passion for change. Executives have to support this with a sense of urgency and the ability to manage resource trade-offs to support the changes. Leadership at the middle level of management must assume that the vision of the CEO is shared and that attainable intermediate goals are designed. This level of management must be able to create discernable critical success factors and measures. Teamwork and communication skills must be sharpened. In addition, the ability to manage expectations during each state of the change process is needed, and the techniques for doing this are often weak. Today managers try to get their performance evaluated at specific levels. Meeting system-level goals is not included in incentive systems, and organizations have weak linkages across organizational lines so the capability to measure contributions from different sources does not exist. These ingredients all imply the need for major new investment in the leadership development process if change is to be effectively managed.

Like drawing and art, management and leadership are not best learned in a classroom but in a workplace where one can work under the guidance of one who knows how to do what one wants to learn how to do.

—Russell L. Ackoff, professor at The Wharton School

- Information Systems—Strategies are often implemented with no way to measure their effectiveness. New investment should not be made without adequate tools for evaluation of their performance. Even if the costs are substantial, concurrent plans for upgrading information systems can be critical success factors in watching the "needle" change as progress is achieved. New measures may be needed that evaluate changes in cycle time, improved decision times, and tracking of costs related to reduction in performance gaps. All of these required elements imply considerable investment. The investments are needed in technology, people, and information systems. Change processes that do not have substantial new investments in all of these categories many times fail. It is critical to balance resources and to plan the change process. The achievement of an Organizational Alignment strategy offers the promise of being able to radically change the organization. The change should occur with a concomitant resource investment that is carefully targeted to meeting customer needs and to make the change process deliver results that are important in the marketplace. Supply chain management is frequently cited as an important new catalyst for Organizational Alignment. Because it is comprehensive and because it cuts across organizational lines, Organizational Alignment qualifies to be categorized as a strategy.

Definition: *Supply Chain Management* is the creation of a management process for integrating decisions, plans, and information systems from customer requirements through the fulfillment process to the suppliers of materials.

Components of the process are purchasing, production/manufacturing, distribution, transportation, product handling, and customer service. For some products, this process can also include recycling and disposal. The major advantage of this approach is that because it is comprehensive, there are substantial opportunities to reassess the way that customers obtain real value-added from the organization. If all elements of the value chain are identified and aligned, barriers to meeting and exceeding customer expectations can be identified.

The method of systematically evaluating how the customer obtains value from the supply chain helps develop a catalyst for change. The synergies and creativity that are unleashed have a dramatic effect on an organization. The impact is that breakthrough changes emerge and core processes and core outputs are defined. The supply chain strategy is useful at the beginning of the change process because it creates a blueprint of the way that an organization functions. This blueprint can discern ways that the organization is out of alignment with customer needs, and thus it is useful in the alignment phase of the business transformation process. Another very powerful tool for accomplishing alignment with customer needs is the JIT II process.

Bose Shows the Way

Bose Corporation used a strategy that unleashes substantial synergy by focusing on the suppliers' hidden pool of talent, knowledge, and resources. Developed at Bose, this practice helps empower suppliers. The supplier acts as an in-house change agent and delivers new value by seamlessly integrating the materials flow process and aligning delivery schedules. The alignment of the supplier organization with the manufacturing process translates into better concurrent engineering, improved designs, and more opportunities for delivering value to customers. For example, one supplier has added accessories to Bose products that make the products more useful in diverse applications.

Once alignment has been achieved, it is possible to add to the resource mix. Some logistics providers add to capabilities by becoming third-party providers of logistics service. The use of this pool of highly specialized technology, information systems, and skilled logistics managers helps an organization add to resources quickly with little new investment. Capabilities to meet customer needs are enhanced. Therefore this approach is useful at the mobilization phase of the change process.

Regardless of the strategy employed, measures of success need to be established early. Also, linkages to a continuous improvement process, such as value analysis and value engineering, are important as a means for institutionalizing the change process. The benefits from use of Organizational Alignment range all the way from Federal Express and the creation of a new industry definition to the use of in-plants that reduce operating costs and generate new ideas for products at Bose Corporation. Xerox used supply chain management and discovered assets of $1.3 billion that could be sold and converted into immediate profits.

Alignment Tool #2 is a form that can be used to help facilitate the goal development process. Once the goals are finalized, this form should be completed for each goal and distributed or published throughout the organization. This then leads to the 3- to 5-year integrated business plan that delineates what has to be accomplished and by when.

You are the CEO and you need to choose a new strategy for change. Which one will work for you? The strategies described in this book can help you chart a course that meets the board's demands. Maybe there is a way to save the organization, keep your job, and avoid the headhunters after all!

ALIGNMENT TOOL #2: GOAL STATEMENT FORM

We find using the following form provides a useful input.

Instructions: Please read each statement and indicate the extent to which it describes the reality in your organization as a whole. Your responses should reflect what you have experienced as well as what you have generally observed.

Rate each statement using a 10-point scale. The left side of the scale indicates that you *totally disagree* and the right side that you *strongly agree*. If you do not know, indicate so next to the question.

- Goal:
- KPI aligned to:
- Time frame:
- Goal meaning/Definition:
- How will it be measured?
- How will it impact the organization in 1 year? In 3 to 5 years?
- How will it impact our external customers and our internal customers?
- What is management's role?

A well-defined vision and commitment to it produces focused results.

HJH

2

Alignment Considerations

A plan is only as good as the people who implement it.

—HJH

VISION-CENTERED ORGANIZATIONAL ALIGNMENT MODELS

Most executives spend their efforts trying to reduce short-term costs and increase short-term profitability. They concentrate on targeted savings of 5% or less and usually develop these savings in one location without consideration of the long-term consequences of the savings. This is Micro-Alignment. On the other hand, when the scope of the change crosses organizational lines, there is a radical change in the nature of the product, or service, and these changes have the potential for developing major breakthroughs (e.g., 50% savings in cost or time), they are Macro-Alignment changes.

A systematic approach for identifying the opportunities for alignment changes and using them to gain corporate competitive advantage is called an Organizational Alignment System. Most organizations have adopted a partial approach to alignment management. One of this book's objectives is to accelerate the process by which organizations consider this option for corporate development. In this chapter, we provide an overview of the components of Organizational Alignment Systems and a rationale for using successful case studies as potential elements of a more comprehensive approach for the management of change and development of competitive alignment strategies.

Approach

The Organizational Alignment System (OAS) is a basic change management model that is loosely based upon an integration of the Rummler-Brache process and the Harrington Institute models. There can be no development of new strategy or implementation of strategy without a logical approach to change management. Organizations whose goals and values are misaligned with its culture will find behavioral change difficult. Despite the presence of a unifying vision, the values and the purpose-related mission statements are static. What makes the vision a reality is that the organization change mobilizes and takes a new direction. The new direction needs to have integration and consensus: a consensus that the direction will deliver the expected desired results through capable processes. Managers and executives who are not fully aligned will find resistance to change that dominates the Strategic Planning process and often derails initiatives.

Outcomes

The need for organizations to have a logical framework for aligning their supply chain around the common vision has led to the creation of value chains. These chains describe the logical activities needed to deliver value to customers using materials, suppliers, and manufacturing processes. The need for every activity in the chain to be identified helps locate areas where the organization does not have alignment. As a diagnostic process, the supply chain often can be used to define areas that are out of alignment.

Also, the information needed to track the supply chain and to implement strategic initiatives often implies the need for new information systems. For example, AMOCO spent millions redefining the supply chain, but the new information system that was required to monitor the chain was going to take 5 more years to develop and millions more to build. The manner in which the system is designed and how it functions can be a critical success factor in the development of the Organizational Alignment model. Federal Express could not have offered overnight delivery without a vast investment in alignment through information systems and logistics. These systems often require alignment across organizational lines, common databases, and common definitions. For example, Policy Deployment is severely hampered by inconsistency in the information system management, and this is where Business Rules Analysis can be used to discover and fix information system disconnects that inhibit change.

An insurance company that was consistently chasing and ultimately losing new, large accounts traced the problem back to subjective and noncompetitive strategies used in underwriting new policy quotes. Once this was rectified, expansion became substantially easier to accomplish. Information systems can be screened to determine if leverage opportunities exist. By identifying policy alignment enablers, such as the creative use of interest incentives to promote sales in specific markets, alignment is enhanced and it is much easier to implement change. Included in the definition of information systems is the use of new information technology, such as handheld scanners, computers, and bar-coding technology. Without rapid deployment of new technology, the transmission rate of data needed to support improvement is not feasible, efficient, or effective.

Key Implementation Challenges

To visualize how an Organizational Alignment program works, we first need to view the organization as a system. Organizations are dynamic systems and like all other systems, they function best when their components are designed to work together smoothly and efficiently. Any change we introduce to an organization must be aligned to fit the existing system or to modify the system to accept the change.

Organizational Alignment requires compatibility, mobilization, and integration between the strategic and cultural "paths," as well as a fundamental consistency within them. Values should be compatible with goals and a work group that values flexibility should think twice about goals focused on developing very tight control systems. Day-to-day behavior should be consistent with stated values: a group that values responsiveness should not answer customer requests with "it's not my job."

Organizations have traditionally emphasized the strategic path over the cultural change side of the equation. Most invest considerable effort in defining strategic goals and objectives. Fewer mobilize the cultural path with clearly defined statements of values, and fewer still make a consistent effort to ensure that values and strategy are integrated and compatible and that work behavior represents their values. This is in spite of the fact that the way we do things influences results fully as much as what we do.

Organizational values, like organizational goals, are an organization's necessities. Maintaining an aligned organization requires clarity about values and their vision alignment, as well as strategies and goals. For example, achieving and maintaining market share requires setting relevant

Stop Doing What Comes Naturally	Reinvent Its Future
Slow down, panic	Speed up, stay cool
Wait for instructions	Take the initiative
Get ready for action	Get going, act now!
Try harder at more of the same	Shoot for the impossible, the absurd, the breakthrough
Try not to harm or break things	Welcome controlled restructuring
Avoid risks, mistakes	Welcome prudent risks, corrects mistakes once and for all
Benchmark others' success	Aim for perfection, Six Sigma
Be loyal to and improve the "as-is"	Be the devil's advocate, stir the pot
Accept the nature of the business	Have faith in realignment
Can't predict the future	Help shape and define the future and make it happen
Behave like adults in moderation	Act like children with enthusiasm and passion for the unknown

FIGURE 2.1
The change that is needed.

goals and testing actions and decisions against those goals. It also requires communicating relevant organizational values and ensuring that typical behavior in the organization reflects those values. Figure 2.1 is an example of how one organization needed to change.

Gauging Effectiveness

In recent years, increasing competition and rapid change have generated more interest in the values side of the organization, which is the component most strongly associated with culture. When customers perceive less and less differentiation among organizations based upon their products and services, they begin to place more and more importance on how those organizations work with them. The key focus of the aligned leader is the building of trust in the organization or team she or he is leading. Trust building is a task that requires an enormous amount of energy and time and revolves around six key competencies: attitude, collaboration, joint accountability, acknowledgment, commitment, and cooperation.

ALIGNMENT TOOL #3: VISION-CENTEREDNESS DIAGNOSTIC SURVEY

We find the following survey useful.

Instructions: Please read each statement and indicate the extent to which it describes the reality in your organization as a whole. Your responses should reflect what you have experienced as well as what you have generally observed.

Rate each statement using a 10-point scale. The left side of the scale indicates that you *totally disagree* and the right side that you *strongly agree*. If you do not know, indicate so next to the statement.

Please take the time to respond to the open-ended questions at the survey end, as your responses are critical to improving the readiness for change. Be honest in your responses as there is no right or wrong answer. Your responses will remain completely confidential.

1. We can explain the Vision we are driving toward and explain it in 5 minutes or less.
2. We can describe the behaviors that our leaders want to witness more often.
3. We understand how the Vision affects our jobs, and we are motivated by the Vision.
4. We are confident that projects that are inconsistent with the Vision will be terminated.
5. We believe that achieving our Vision is feasible and will make us a better organization.
6. Employees and leaders believe that achieving the Vision in the time frames given is feasible.
7. Our Vision takes into consideration our organizational capabilities, the market environment, and the competitive trends.
8. All employees are held accountable for annual performance toward the Vision.
9. Employees feel that our Vision is flexible enough that it can be changed in light of changing conditions and feedback from customers and employees.
10. Employees believe that our organization has the skills necessary to achieve the Vision, that it is backed by sensible actions, and they understand how these actions will help achieve the Vision.

Totals: Add up the scores for each of the 10 statements: _____

Note: Grand Total: A total of 50 or less indicates very serious problems and 80 to 100 indicates no serious issues, while 50 to 80 indicates need for improvement.

Answer these questions:

1. Explain how senior leaders communicate and deploy performance expectations.
2. Explain how well senior leaders establish and reinforce an environment for individual initiative.
3. Explain how your organization's education and training support the achievement of its objectives; build employee knowledge, skills, and capabilities; and contribute to improved performance.
4. Explain to what extent the organizational goals, strategies, action plans, and performance measures are aligned to achieve overall organizational alignment and involvement.

MACRO-ALIGNMENT/MICRO-ALIGNMENT AND CULTURE CHANGE: KEY COMPONENTS

Most executives spend their efforts trying to reduce short-term costs and increase short-term profitability. They concentrate on targeted savings of 5% or less and usually develop these savings in one location without consideration of the long-term consequences of the savings. This is Micro-Alignment. On the other hand, when the scope of the change crosses organizational lines, there is a radical change in the nature of the product, or service, and these changes have the potential for developing major breakthroughs (e.g., 50% savings in cost or time), they are Macro-Alignment changes. In the midst of all this is culture, which surrounds the organization like an invisible gas and which causes people to behave in a specific manner. This invisible gas is impervious to change and thrives in a state of status quo. A feeling of urgency greatly helps in putting together the right group of people to guide the alignment effort and in creating the essential teamwork within the group. Most people are willing to pull together, even if there is no short-term personal gain, even if risks are involved. But additional effort is needed to get the right people in place with the trust, commitment, and teamwork to do the job. Mission and vision represent long-term organizational intent. They provide guidance about organizational purposes and a picture of the organization's future performance (vision). For example, here is a mission statement for a hypothetical financial services organization: "We provide products and services to business customers that help them make well-informed, timely financial decisions." Accompanying that mission statement, or as part of it, might be a vision of its impact: "We see our customers developing a well-founded confidence in their financial decisions and an increasing security about their financial futures." Such statements provide general guidance to everyone in the organization in making choices about strategies, customers and markets, products, and services.

Goals and Values

Goals and values provide greater direction about where the organization is going and by what means. They establish how the organization intends to allocate resources to accomplish the mission/vision over time (goals) and how it intends to behave as it does so (values). For example, supporting the mission and vision above might be strategic decisions or goals like these:

- To provide a full line of financial services to small and mid-size organizations.
- To gain a competitive advantage through top-quality customer service.

Statements like this give people guidance about how to allocate resources, and where to invest their time and effort. In addition, the organization can make statements about the kinds of values it considers important, such as the following:

- Partnering: We work in partnership with our customers, freely sharing information, ideas, and plans.
- Initiative: We encourage people at all levels to take initiative to meet customer needs, and we support them in doing so.

Statements like these provide guidelines for how people are expected to behave in working with customers and how managers are expected to behave toward customer support personnel. Mission/vision, value, and strategy statements tell people "what we are about" and guide members of the organization in setting priorities and choosing how to behave.

Objectives and practices are the institutionalization of strategies and values. They represent decisions about how to implement those strategies and values: the objectives people set for themselves and the results they expect of their work units; the typical ways they interact with customers and others both within and outside the organization. For example, managers might support a goal to increase market share among small and mid-size organizations by setting specific sales objectives for those markets, or by setting product development objectives around the needs of small and mid-size customers.

> If the corporation's goal is to capture and maintain increased market share, profitability and shareholder return, the new venture must contribute significantly to that goal.
>
> **—Joey Tamer,** *Handbook of Business Strategy,* **2005**

Managers might support a value of partnership by practices like holding regular meetings with clients. They might support a value of initiative by practices like giving frontline customer service personnel the resources and authority to take independent action in meeting customer needs.

Activities and behaviors are the execution of intent—the ultimate determinants of organizational performance. These represent what really happens in an organization on a day-to-day basis: the activities people choose to invest their time in and the way they behave as they perform those activities. Statements of mission and vision, values, and strategies are meaningful only insofar as they are translated into action. For example, a strategic decision to build a competitive edge through customer service becomes reality when people throughout the organization engage in activities like acting on customer feedback and testing decisions made anywhere in the organization for their potential impact on customers.

Values of partnership and initiative become reality when people engage in behaviors like inviting personnel from other groups to planning meetings and taking action to meet needs as they arise, rather than waiting for approval.

Results Are What We Are After

Results are the outcomes an organization produces as a function of the activities and behaviors performed. They can be measured in a variety of ways: financial indicators, product and service measures, customer retention rates, sales measures, employee and customer attitude surveys, measures of market share, and so forth. The way an organization chooses to measure its performance determines its ability to stay on track—to evaluate its progress against values and strategic goals.

The stock market's myopic focus on this quarter's earnings thus penalizes those firms that are currently investing heavily in their people, relative to others.

—Laurie Bassi and Daniel McMurrer, *Handbook of Business Strategy*, 2005

For example, an organization that measures results exclusively in terms of outcomes like sales volume and profit will have a pretty good picture of short-term success but will be missing information that may be critical to long-term health, such as customer retention measures. The strategic and cultural "paths" do not operate in isolation. They interact with the organization's external environment, with its internal support systems, and with its stakeholders.

- External Environment: This includes a host of factors, such as the economy, sociopolitical environment, competition, governmental policies and regulations, and the state of the technology. Any or all may influence an organization's strategy or values. For example, heavy competition in the large corporate market, and the costs required to penetrate it, may influence an organization's decision to concentrate on the small and mid-size business market. Increasing evidence of an organization's impact on the physical environment may result in placing greater importance on social responsibility as a value.
- Stakeholder Value: Stakeholders include any group that is significantly affected by the organization's performance, such as customers, shareholders, suppliers, management, employees, special interest groups, and even the general public. These groups have different relationships with, and expectations of, the organization; understanding these expectations is a key factor in organizational decision making. For example, whereas shareholders and financial analysts may judge an organization heavily in terms of its growth or profits, customers may be making their evaluations on such factors as responsiveness, quality and range of services, or environmental sensitivity. Organizations need to take both sets of expectations into account.
- Support: Leadership and systems function as "performance levers" that help (or hinder) people in implementing strategies and values and producing results.

Leadership reflects the ability of leaders and managers to focus on the "big picture" and to serve as both models and coaches in support of strategies and values. An organization that values respect, for example, requires leaders who model respectful treatment of others. An organization that wants to be known for innovative, cutting-edge products requires managers who support experimentation and calculated risk-taking and are willing to accept the inevitable failures that risk-taking entails.

Organizational systems include reward systems; information systems; performance appraisal, compensation, and benefit systems; organizational structure and reporting relationships; training and development; work design; and administrative policies. Compensation systems for salespeople that focus exclusively on revenue targets can create pressure to violate values about treatment of customers, or to ignore Strategic Plans for penetrating selected markets. Similarly, centralized control policies designed to ensure consistency can get in the way of responding to customer needs unless those policies are flexible and balanced by reward systems or other factors that support responsiveness to customers.

Organizational Alignment occurs when strategic goals and cultural values are mutually supportive, and when key components of an organization are linked and compatible with each other. Market strategies should be consistent with organizational values and should be so perceived by members of the organization. Group objectives should be derived from organizational strategy and supported by management practices. People's day-to-day activities and behaviors should be consistent with mission, strategy, and values. Organizational systems and leadership should support those activities and behaviors.

Key Implementation Challenges

Few organizations will achieve "complete" alignment. The real magic comes from using alignment in combination with the existing architecture coupled with a new generation of data modeling and performance integration forecasting tools. Organizations now have the ability to access capabilities from their own systems and from their partner organizations to create the building blocks of aligned organizational performance processes that will meet the needs of each customer in each market. An organization can then design and deploy programs that require the involvement of people and systems, within and across a business's boundaries.

An Organizational Alignment project may begin with an organization's needs assessment, followed by an organization-wide alignment analysis to recommend and implement appropriate interventions. More often, it is a companion effort, accompanying a major change that has organization-wide implications. For example, here are a few situations in which Organizational Alignment models and methods have been applied:

- An organization conducted a major reengineering effort, including skills training, but found that people were often not following new procedures and the organization was not realizing the anticipated gains. An alignment effort helped to modify organizational culture and systems to support the new processes.
- A company formed by a consortium of 10 different organizations, representing 10 different nationalities, found itself significantly behind schedule and over budget. Under the direction of its CEO, people worked to create a core culture that people from all nationalities could buy into and that supported the venture's direction.
- Engaging in benchmarking efforts, capturing and sharing best practices, intensively partnering with important customers, holding focus groups, and keeping open lines of continuous two-way sharing of information with other external constituencies are important feedback mechanisms. Finally, the explicit creation of listening posts at various nodes in the organization can enhance the process of learning from change. Many of the coordination vehicles, steering committees, support teams, advisory groups, cross-business and cross-functional councils, and communities of practice can also double as listening posts.

These important feedback mechanisms seldom emerge on their own during transformations, unless there is a crisis situation. Basically, organizations lack a nervous system, so transformational leaders have to create and nurture feedback loops to facilitate organizational learning. Transformational leaders should be able to anticipate needed developments along a corporate transformation path and not be overwhelmed by firefighting or missed opportunities along the way. To that end, they put in place a network of active feedback mechanisms early in the process, and make sure that their channels stay wide open and unfiltered in order that the cultural change aspects of the Organizational Alignment reach their true value level.*

> If decision makers overlook uncertainty and behave as conventional risk minimizers, their organizations are likely to miss opportunities.

> **—Mehmet Ugur, University of Greenwich**

Gauging Effectiveness

These and similar problems have been addressed through an Organizational Alignment effort to ensure that components of the organizational system are working together in a way that effectively meets an organization's needs. Organizational Alignment is a business discipline that deals with both operational processes and employee behavior on a systemic, outcome-focused basis. Because it focuses on meaningful results and an organization's drivers, it is often readily seen as relevant by an organization's management.

Your Organizational Alignment teams are operating effectively, with clear goals and trust among their members, if they consistently do the following:

- Stay on track regarding their approach and direction.
- Rarely need to revisit completed work.
- Have open and honest discussions about problems, issues, and progress.
- Use conflicts constructively.
- Make tough decisions rapidly.
- Communicate frequently and clearly.
- Are effective at resolving issues using all members' knowledge and input.
- Work in and engender a relaxed and enjoyable atmosphere.
- Allow members to function autonomously, are clear about their roles and responsibilities, and integrate well with the other guiding teams.
- Understand and believe in the importance and urgency of the change effort.

---

* Andrew Pettigrew and other experts have made the point that if the cultural change aspects of realignment do not reach optimum value levels (including shareholder value), the new culture will remain fragile and vulnerable, especially when the strong leader who created the change moves on or retires. His studies included research from Eskimo villages covering a 100-year period and their efforts to realign their cultures to reinforce learnings and optimize their environment. See John Kotter and James Heskett, *Corporate Culture and Performance,* Chapter 6, for an interesting treatment of this topic; also see an interesting historical treatment by Franz Boas, "The Central Eskimo," Bureau of Ethnology Annual Report No. 6 (Washington, DC: Smithsonian Institution, 1884).

If any of the following warning signs appear more than once, it is an indication that some or all of the teams do not have the necessary skills and attributes, are not senior enough in the organization, or do not have the necessary respect of the stakeholders to lead the change effort successfully. Consider yourself warned if your teams do the following:

- Cannot get the resources, information, and support to progress rapidly
- Do not motivate and inspire others to participate in the change
- Cannot gain the support and help of specific groups in the organization
- Cannot get on the leadership's meeting agendas
- Cannot get one-on-one time with key leaders when needed
- Constantly need to consult experts to make decisions
- Cannot make important decisions without a lengthy approval and review process
- Do not have the confidence of other senior staff

If a competent key senior leader cannot mobilize a design team with the right mix of credibility, expertise, and leadership, it is a sure sign that the organization has little appetite for the change proposed. For the organization to release its best people for a change initiative, it has to share the need and urgency for the change. Gauging effectiveness at this activity in the change process merits a different type of diagnostic tool: one to determine how the members of the guiding teams see themselves and also to determine how the stakeholders in the change process view the guiding teams.*

Every good change initiative needs to be driven by groups of influential, effective leaders. These groups, the design and development teams, help the organization understand why the change is needed and thus must be fully committed to the change initiative, be fully respected within the organization, and have power and influence.† This team self-assessment tool will help determine how leaders see their own behavior as guiding team members and how that perception influences team success and progress toward defined goals.

---

* For example, in the auto industry, design teams usually made two different clay models of a car to be submitted for final approval by top management. There have been occasions when the president liked one model and the chairman liked another. Designers would have to compromise and combine disparate parts of the two models, after a further series of diagnostics was performed.

† David Nadler and Michael Tushman have called this the "many bullets" approach in their important work, "Organizational Frame Bending: Principles for Managing Realignment and Reorientation," *Academy of Management Executive* 3, no. 3 (1989): 194–204.

ORGANIZATIONAL ALIGNMENT READINESS AND CULTURE

We encounter organizational cultures all the time. When the cultures are not our own, their most visible and unusual qualities seem, at the very least, striking: the look of the traditionally dressed IBM salesperson or Wall Street executive, the informality and high-tech look of Apple and many other technology companies, the overnight speedy delivery promise of Federal Express, who were pioneers in the use of new technologies to deliver on their goals of consistency and speed, while at the same time systematically pursuing continuous improvement and alignment in all of their logistics processes. FedEx is a worthwhile benchmark organization that effectively models the application of Organizational Alignment systems that have included serious culture change outcomes. It is no wonder that they helped define a new industry and that they were the first major service organization to receive the Malcolm Baldrige award for performance management.

When the cultures are our own, they often go unnoticed—until we try to implement a new strategy or program that is incompatible with their central norms and values. Then we observe first-hand the power of culture. The term "culture" originally comes from social anthropology. Late 19th- and early 20th-century studies of "primitive" societies such as Eskimo, South Sea, African, and Native American revealed ways of life that were not only different from the more technologically advanced parts of America and Europe but were often very different among themselves. The concept of culture was thus coined to represent, in a very broad and holistic sense, the qualities of any specific human group that carry forward from one generation to the next. The *American Heritage Dictionary* defines "culture," more formally, as "the totality of socially transmitted behavior patterns, arts, beliefs, institutions, and all other products of human work and thought characteristics of a community or population."

We have found it helpful to think of organizational culture as having two levels, which differ in terms of their visibility and their resistance to change.* At the deeper and less visible level, culture refers to values that

---

* Because the word "culture" is used with many different meanings in everyday writings, talk, and conversations, we have chosen the meaning as a combination of shared values and group behavior norms, as reflected and described by Edgar Schein in his book *Organizational Culture and Leadership* (San Francisco: Jossey-Bass, 1985).

are shared by the people in a group and that tend to persist over time even when group membership changes. These notions about what is important in life can vary greatly in different organizations: in some settings people care deeply about money, and in others people care more about technological innovation or employee well-being. At this level, culture can be extremely difficult to change, in part because group members are often unaware of many of the values that bind them together.

> Study results were based on various surveys and concluded that an extreme level of rudeness is rampant in the U.S. workplace that both damages mental health and lowers productivity.
>
> **—Joan Curtice, *Handbook of Business Strategy,* 2005***

At the more visible level, culture represents the behavior patterns or style of an organization that new employees are automatically encouraged to follow by their fellow employees. We say, for example, that people in one group have for many years been regarded as hard workers, those in another group are known to be very friendly to strangers, and those in a third always wear ultra-conservative clothes. Culture, in this sense, is still tough to change, but not nearly as difficult as at the level of basic values.[†]

The measurement system can provide you with a quick indication about the organizational culture, as Figure 2.2 indicates.

---

* Joan Curtice (2005) Want to motivate your employees? Keep your company safe and you will, *Handbook of Business Strategy,* 6:1, 205–208. This paper examines the value of a work environment that is safe, dignified, and respectful. Not only does the law now mandate such an environment, but the author argues it makes good business sense, since typically the results are increased motivation and productivity. The underlying premise is that successful companies practice safety and fairness procedures, and consistently reap rewards for doing so. This article covers the basics of safe work atmospheres, citing examples of the significant costs to firms that do not follow these practices. One of its conclusions is that people are at their most productive when they are not distracted by concerns for their safety or well-being.

† Each level of culture has a natural tendency to influence the other and is perhaps most obvious in terms of shared values influencing a group's behavior—a commitment to customers, for example, influencing how quickly individuals tend to respond to customer complaints. But causality can flow in the other direction too: behavior and practices can influence values. When employees who have never had any contact with the marketplace begin to interact with customers and their problems and needs, they often begin to value the interests of customers more highly. Conceptualized in this manner, culture in a business enterprise is not the same as a firm's "strategy" or "structure," although these terms (and others such as "vision" or "mission") are sometimes used almost interchangeably because they can all play an important part, along with the competitive and regulatory environment, in shaping people's behavior.

Type of Culture	Key Measurements
Financially driven	ROI Service Costs ROA
Quality Driven	Customer Satisfaction Poor-Quality Costs Customer Complaints
Resource Driven	VAE Inventory Costs Cycle Time
Investor Driven	Market Share Stock Prices Profits

FIGURE 2.2

Culture-driven key measurements. ROI, return on investment; ROA, return on assets; VAE, value-added per employee.

Who Is Doing It?

Although Federal Express embodies many of the principles of Organizational Alignment, the company does not have a consistent change model in place and still has not utilized or integrated many of the advanced techniques. For example, an early failure to expand in the European market can be traced to the lack of a consistent approach to change and a failure to recognize the inevitable disconnects that hamper expansion in new countries and with new cultures.

Organizations such as Burlington Railroad, Hitachi, Eastman Chemical, IKEA, and Chrysler provide examples of how to develop a unified approach aligned to new strategic initiatives. One important distinction is that when Japanese organizations change, they have an underlying cultural consensus-building process that supports the achievement of alignment without having to completely reconstruct all of the business models involved. Other organizations may assume alignment and try to mobilize and then discover that they are not capable of achieving their goals. IBM has extremely sophisticated systems for alignment. However, the real question for any organization remains, is the organization's structure aligned with the processes the organization is using?

One of the more productive ways to achieve real progress in Organizational Alignment is in the use of Just-in-Time (JIT) principles. Just-in-Time (JIT) is an inventory strategy implemented to improve the return on investment of an organization by reducing in-process inventory and its associated carrying costs. To achieve JIT, the process must have signals of what is going on elsewhere within the process. This means that the process is often driven by a series of signals that tell production processes when to make the next part. Quick communication of the consumption of old stock, which triggers new stock to be ordered, is key to JIT and inventory reduction. This saves warehouse space and costs.

When organizations rework their inventory practices, they inevitably discover inefficiencies that need to be eliminated for progress on strategic initiatives to occur. In the case of Japanese organizations such as Toyota, they deploy all their resources, including supplier resources, in an integrated fashion to achieve breakthroughs in product cycle time reductions and in cost savings due to the elimination of inventory expense. Once the JIT process savings are obtained, the question remains how to continuously upgrade this process. JIT II can provide part of the answer.

The JIT II technique provides an ongoing dialogue for continuous improvement and Organizational Alignment with suppliers that can deliver hidden competitive advantages to an organization. These advantages were trapped by the fact that communication processes did not move across organizational lines and with suppliers because of bureaucracy. The JIT II process captures these opportunities for product improvements cost savings, and inventory management. It allows a new culture to form around the JIT II "in-plants." The in-plant is an agent provocateur or catalyst for change that, in many ways, can revitalize the change process. It is therefore an integral technique to developing Organizational Alignment across the supply chain. One of the pioneering applications of JIT II is at Bell Labs. Many other organizations, such as IBM, Intel, Ford, and Honeywell, also are taking advantage of this technique. Unfortunately, many organizations use JIT or JIT II as only their logistics strategy. They have mobilized without alignment or integration.

Outcomes

Although the aligned supplier team is fundamental to many organizations, in some cases such as Whirlpool and Hewlett Packard, the use of internal cross-functional teams has helped generate opportunities for value-based strategic initiatives. Customer data and team meetings can be used to innovate across the supply chain. Furthermore, when Organizational Alignment has shifted from an internal focus on logistics and cost-related improvements that are internal to an external customer-phased process, it is a value chain.

Organizational Alignment strategies easily emerge from the use of customer definitions of value. These concepts can be used to redefine the corporate goals and are therefore useful in aligning change. Again, the definition of a value-based initiative does not imply that there is alignment and there may not be much acceptance of the resulting initiatives without alignment. Creating mobilized resources such as supplier partners, cross-functional, externally oriented teams, or Keiretsu-like new corporate networks, is a useful tactic, but it does not form the sole basis of strategic change. Examples of these approaches are contained in later chapters.

Key Challenges

The most ambitious organizations, such as Motorola and General Electric, are determined to use the latest methods to develop competitive advantage. When the past chairman of GE, John Welch, said that he wanted his organization to be World Class, there emerged a need to define standards that exceeded those of competitors. Once change is desired, the question is what form of change will it take to overcome and stay ahead of competition? Later in this book we describe how these organizations use the benchmarking technique to obtain data on these incremental changes and how to implement change.

Benchmarking is a useful data collection process for the Organizational Alignment team and it helps define the goals that will be World Class. Competitive advantage is not valuable unless it enables the organization to leapfrog and stay significantly ahead of competition. In the case of Motorola, the use of a Six Sigma standard for defect reduction is considered World Class because it moved Motorola into a category of virtually defect-free production. Although benchmarking is not a sufficient technique for change, it is necessary to define the kind of breakthrough change that would have a significant competitive advantage. Organizations, such as General Electric and IBM, often use this approach to maintain leadership in their industry and to continually distance themselves from competition.

> Definition: *Business Process Improvement (BPI)* covers the breakthrough improvement approaches process redesign, process reengineering, and process benchmarking.

It is unlikely that an organization can become World Class without the last phase of the alignment process: integration. Organizations, such as Xerox and IBM, are now engaged in a major integration process using the latest techniques of Business Process Improvement (BPI) to redefine key processes that they use for competitive advantage. Just as a car needs a tuned-up engine and a full tank of gas, it also needs to have gauges that can measure progress. The design and use of measures, such as balanced scorecards, helps facilitate change by creating an environment of measurement of progress. If continuous improvement is combined with a measurement system and World Class goals are defined, the car can proceed at an accelerated pace. Progress toward achievement of goals is hampered by the lack of an evaluation-based system. Process efficiency and cost-only scorecards do not work. It is also necessary to evaluate the impact of change on customer satisfaction, quality, and speed.

In later chapters we describe the process of developing an Organizational Alignment strategy. In the beginning, organizations can have as a strategic goal the use of alignment to gain competitive advantage. This is the case of Federal Express. Organizations can develop a systematic change process using the change model in this book. To create a comprehensive master plan containing all the necessary elements of this strategy, it is necessary to conduct a diagnostic assessment of the existing initiatives, perhaps using benchmarking to gain information and to formulate World Class goals.

Resources are mobilized using techniques such as JIT II and supplier partnering. Integration then becomes the major challenge. Also, the alignment of the continuous improvement process, Business Process Improvement, area activity analysis, and Business Rules Analysis become major techniques for achieving integration. When cross-functional teams are used in an adaptive learning process, new information is continuously obtained as the result of solving problems addressed by the teams. Adaptive engineering becomes the major diagnostic process for continuously analyzing the results developed by the cross-functional teams.

However, management must identify disconnects in alignment mobilization shortfalls and areas where more integration is needed. The adaptive learning philosophy, when combined with Policy Deployment or Area Activity Analysis and Quality Function Deployment (QFD), enable the integration toward World Class goals becoming a reality. When these goals involve the need to develop a breakthrough and there are alignment strategies involved, adaptive engineering tools like Business Process

Improvement and Policy Deployment can become part of the engine for change. Whether adaptive engineering is used or not, if breakthroughs in alignments occur, they are likely to provide a major competitive advantage. Organizations can use these concepts to develop a catalyst for change. The catalyst can be the use of the alignment model involving only one initiative, or it can be the entire Organizational Alignment management process.

Lasting corporate change is extremely difficult to accomplish. As a manager or executive, you are trying to change the inertia of a moving aircraft carrier armed only with an idea. Hopefully, the use of the Organizational Alignment framework will make the process of achieving fundamental and significant change possible and executable. We are optimistic that many additional organizations will see these benefits and adopt these concepts either partially or wholly. We are interested in developing dialogues with those who are interested in a more comprehensive and systematic approach to organizational change, and we hope you will join these pioneering organizations and take advantage of one of the most powerful new tools of the 21st century!

ORGANIZATIONAL ALIGNMENT AND E-BUSINESS STRATEGY

Over the past several years there has been a lot of leading edge thinking and many definitions put forth of how technology will impact Organizational Alignment. These definitions are compatible, and debate over which is proper is not time well spent for most organizations. Regardless of the definitions applied, what is common among all the successfully aligned organizations that we have worked with and studied over the past 20 years or so is their ability to identify those four or five key areas of strategic focus that are characterized by the following:

- Customers value the benefits that the focus provides.
- Concentration of resources toward being the absolute best in your chosen areas of emphasis will enable you to excel.
- Excellence in these areas will be difficult for competitors to imitate.
- These areas of focus are your organization's capabilities or what you're really good at, not outcome measures like market share, profit margin, and so on.

Approach

The clarity provided by focusing on a few key goals can help set the foundation for dramatic improvements in performance results. For example, years ago, Hewlett-Packard established the lowering of their product failure rate as one of their key business goals. They also coupled this goal with a very clear objective or measurement, with the specific target being a tenfold improvement in results. Their continuing success in a number of rapidly changing markets speaks for itself.

As previously discussed, culture is defined as a set of shared attitudes, values, goals, beliefs, behaviors, and practices that characterize an institute or organization. They are developed and driven by the collective experiences, education, and social background of the people who make up the organization and highly influenced by the management team. Culture is a lot like an iceberg. Only 10% of it is visible—the rest of it is below the water line and you have to work hard to see it in total. The outward cultural behaviors are easy to see, but the attitudes, personalities, personal standards, judgments, ethics, motives, and behaviors are hard to see and even harder to change.

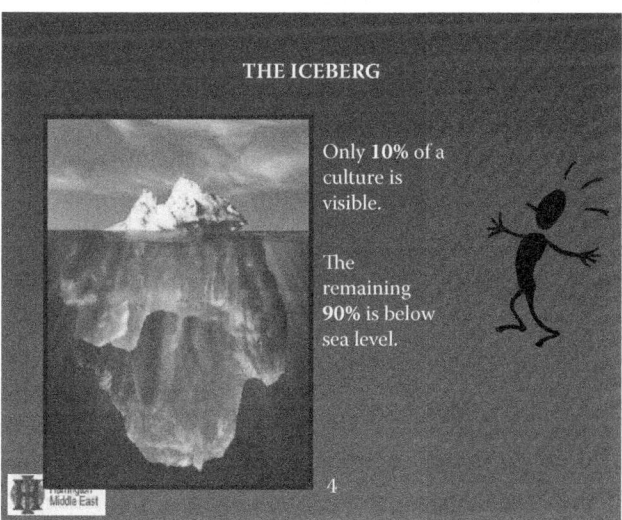

Strategy, on the other hand, can change very rapidly. A new product, a new competitor, or a new acquisition can bring about major changes in the organization's strategy that will require a change in the organization's culture. Organizational Alignment requires that both the strategy and the culture paths be in harmony with each other. Organizations have a tendency to focus on the strategic path and think that the cultural path will take care of itself. Too few organizations ensure that values and strategy are in harmony with actual behaviors and the organizational structure.

> Scenario analysis is a critical tool in any review of the external environment. It is particularly useful in circumstances where it is necessary to take a long-term view of strategy and where there is a high level of uncertainty about external influences.
>
> **—Franc Milburn, journalist**

The organization's mission/vision and its strategy are driven by external factors that make up the organization's environment. Things like technology, competition, government rules and regulations, the economy, and changing customer needs and location drive the strategy. These organizational factors require the final performance results to change in order to meet the organization's stakeholders' requirements. The organization's stakeholders include investors, management, suppliers, employees, employees' families, customers, government, and the community (the general public). It is important to realize that "the way we do things" influences the final results just as much as "what we do."

> At all times, management must know what to do to ensure legal compliance and to create a work atmosphere that is safe, dignified and respectful. Such an environment also ensures that employees can be their most productive.
>
> **—Joan Curtice, *Handbook of Business Strategy*, 2005**

The way management leads the organization is key, as management has the power. Management leadership styles, systems, and organizational structure can either help or hinder the harmony between the strategic and cultural paths. Leadership reflects management's ability to stay focused on the mission and vision and their ability to serve as a positive role model for the employees. It defines the level of risks they are willing to take, the level of participation they are willing to allow, as well as the commitment they have to add value to all stakeholders. It also includes the systems they put in place as well as the organizational structure they developed. This all adds up to the type of culture they create within the organization.

You may ask the question, "When do you need to use the Organizational Alignment methodology?" The answer is that every time there is a change in the Strategic Plan, the cultural side should be considered. This means that each time the organization prepares an organizational plan, its impact on the cultural side should be analyzed. At that point in time it may or may not need to be adjusted based upon the kind and magnitude of the changes in the organizational plan.

Outcomes

It often happens in a flash, right before our eyes: a vast and quick realignment on an evolving electronic commerce (e-business) foundation. It is not just about electronic commerce transactions; it's about redefining old business models, with the support of technology at the core, to enhance and maximize customer value. Organizational Alignment is the overall strategy, and electronic commerce is an extremely important facet of it.

Why is this so-called e-business a big deal in the alignment strategy? CEOs everywhere are faced with shareholder demands for revenue growth and cost reduction, no matter what the organization's environment is. They've already used Lean and Six Sigma, or perhaps BPI, to redesign, downsize, and cut costs. Now CEOs are investigating new strategic initiatives in order to deliver results, and many are looking at using technology to transform their business models. Organizational alignment is being driven by a significant, powerful, and continuously unfolding development: each day, more and more individuals and organizations worldwide are being networked and linked electronically. Although on the surface this may not appear to be earthshaking, digitally bound consumers and organizations in a low-cost way may prove to be as significant as the invention of the cotton gin, electricity, the telephone, and Ford's assembly line all rolled into one.

To understand this concept in greater detail, let's start by looking at one of the rules of e-business alignment: technology is no longer an afterthought in forming organizational business strategy but rather the primary cause and driver. Although the effect technology alignment has on organizational business strategy may not be clear at the onset, it is relentless and cumulative, and it comes in waves. As the ocean erodes the shoreline sometimes rapidly and sometimes slowly, so will technology erode strategies, causing an organization's business model to behave in different and unpredictable ways. Consequently, technology alignment is not something that organizations can ignore, as it poses the most significant challenge to the organization's business model since the advent of the micro-computer itself. Although the computer automates tasks with the promise of increasing organization's business speed, it hasn't yet begun to fundamentally alter the organizational foundation in the way that the Organizational Alignment process does.

Whenever any entity in the value chain begins to do activities electronically, organizations up and down that value chain must align and follow suit, or risk being marginalized and outsourced. Accordingly, rethinking and redesigning the business model is not only one of many options available to management, it is the first activity to profiting and even surviving in the information era. Are executives and managers aware that the impact of these changes is of critical importance? In our experience, some are; many are not. The majority of managers are too busy dealing with a multitude of pressures of day-to-day operational problems. Executives can't afford to think about alignment too much, as they try to get more impact from their current business models. Time is tight, and resources are tighter. If they sit around inventing elegant strategies and then try to execute them through a series of flawless decisions, the current business is doomed. If they don't think about the future, the business is doomed. Catch-22! Finally, to do business differently, managers must learn to see differently. As John Seely Brown, one-time chief scientist at Xerox, put it, "Seeing differently means learning to question the alignment framework through which we view and frame competition, competencies and business models."[*] Maintaining the status quo is not a viable option. Unfortunately, too many organizations develop a pathology of reasoning, learning, and attempting to innovate only in their own comfort zones. The first activity to seeing differently is to understand that all business is about structural alignment transformation.

ALIGNMENT TOOL #4: ORGANIZATIONAL STRATEGY ALIGNMENT SURVEY

We find the following survey useful.

Instructions: Please read each statement and indicate the extent to which it describes the reality in your organization as a whole. Your responses should reflect what you have experienced as well as what you have generally observed.

---

[*] John Seely Brown, ed., *Seeing Differently: Insights on Innovation* (Boston: Harvard Business School Press, 1997).

Rate each statement using a 10-point scale. The left side of the scale indicates that you *totally disagree* and the right side that you *strongly agree*. If you do not know, indicate so next to the statement.

1. Resources are allocated in the most efficient manner and at the lowest cost possible.
2. Processes between suppliers and our organization are often integrated for greater efficiency and speed.
3. Sales intelligence is managed and converted from facts, details, and insights into information we can use.
4. Our organization is dedicated to measurement systems that monitor and measure all processes.
5. Our organization is continually looking for ways to reduce costs and improve both service and quality.
6. We manage customer expectations and provide a manageable set of product and service options.
7. Our business strategy takes into consideration our organizational capabilities, the market environment, and the competitive trends.
8. Our business strategy formation process includes employees who are held accountable for annual performance toward the goals and objectives.
9. Employees feel that our business strategy is flexible enough that it can be changed in light of changing conditions and feedback from customers and employees.
10. Employees believe that our organization has the Strategic Planning skills necessary to achieve the Vision, that it is backed by sensible actions, and they understand how these actions will help achieve organizational objectives.

Totals: Add up the scores for each of the 10 statements: _____
Note: Grand Total of 50 or less indicates very serious problems and 80 to 100 indicates no serious issues, while 50 to 80 indicates need for improvement.

3

The Organizational
Alignment Methodology

Organizational Alignment occurs by design and follow-through. Without it, the outcome is hit or miss with a lot more misses than hits.

—HJH

THE ORGANIZATIONAL ALIGNMENT CYCLE

Organizations that have been successful, based in part on good planning, know that they must provide the organization and its employees with a road map to help "translate" the vision and mission into "things people can do." The next most critical element of what's in a good plan is the strategic focus on the organization in terms of "how" it will compete.

The Organizational Alignment Cycle (OAC) is divided into six phases:

- Phase I. Strategic Plan—This defines directions and sets expectations.
- Phase II. Processes and Networks Design—It defines the way activities are joined together and the way information is distributed.
- Phase III. Organizational Structure Design—This defines the responsibilities and decision-making power within the organization.
- Phase IV. Staffing—This defines the type of people who are needed and the skills they must have.
- Phase V. Rewards and Recognition System Design—This is used to motivate and encourage people to behave in the desired behavior pattern.
- Phase VI. Implementation—These are the processes and activities that are required to bring about the desired changes in the organization.

It includes preparing the procedures and policies that allow the organization to operate at maximum efficiency.

The Organizational Alignment Cycle (OAC) basically says that an organization should align six elements to achieve a sound organizational design: strategy, processes, structure, people, rewards, and implementation as illustrated in Figure 3.1.

- Phase I. Strategic Plan—This phase specifies the goals and objectives to be achieved as well as the values and missions to be pursued; it sets out the basic direction of the organization. It includes two different plans:
  - Strategic Business Plan (SBP)
  - Strategic Improvement Plan (SIP)
- Phase II. Processes and Networks Design—This phase outlines the functioning of an organization and the circulation of information, how outputs are generated.
- Phase III. Organizational Structure Design—Through the organizational and the departmental design, this phase determines the placement of power and authority in the organization.

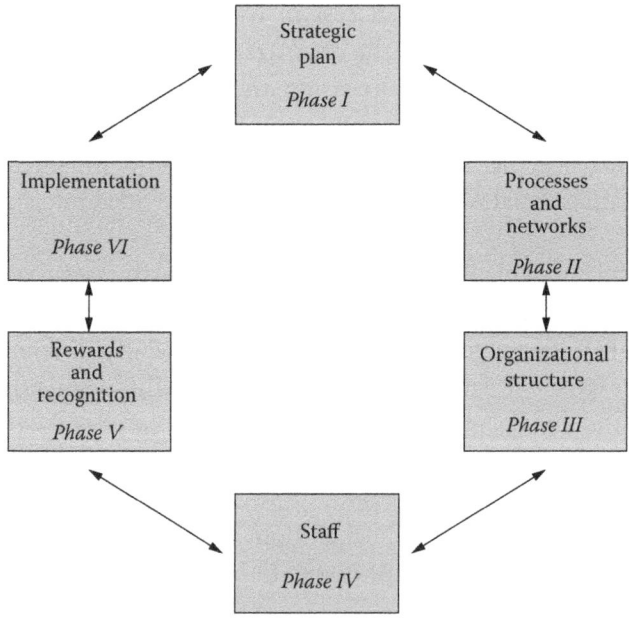

FIGURE 3.1
Organizational Alignment Cycle.

- Phase IV. Staffing Phase—This phase outlines the skills and mind-sets required by the strategy and structure of the organization.
- Phase V. Rewards and Recognition—This phase aligns the goals of the employee with the goals of the organization by providing motivation and incentives for the completion of the organization's strategic direction.
- Phase VI. Implementation—This phase defines how the changes are installed into the organization.

Each of these six phases will be presented in detail later in this book.

Typical OAC Application

We illustrate this alignment concept by applying the OAC to the organizational implementation consequences of EFG International's (EFG) business model (See Figure 3.2). EFG is a private bank that was founded in 1995 by Jean-Pierre Cuoni (with major financing from the Latsis family) and has grown into a bank that currently has almost 90 billion SFr. of assets under management.* Its business model is entirely built around EFG's bankers, who are called client relationship officers (CROs). Bankers run their own P&L based on the business they get from their clients, and they are remunerated in consequence. This stands in contrast to the discretionary bonuses that most relationship managers received in other banks and which are decided by their superior managers. Everything at EFG is designed for and around CROs, and they have the liberty to decide who they want as clients and how they want to serve them, within some general guidelines.

> Organizational Alignment is a process, not disjointed different activities like most organizations treat it.
>
> —HJH

---

* Alex Osterwilder, Bernett & Partner, *Rediscovering the Star Model*, NY, 2008. When the Arvetica consulting organization organized a mini-workshop on private banking business models with a former banker and a local headhunter, one of the participants brought up an interesting idea to help banks align their organizational design and their business model. The basic idea consisted of using the Star Model concept originally developed by Jay Galbraith (and amplified by Jim Harrington) to achieve a strategic fit between an organization's business model and its organizational design. The Star Model basically says that a company should align five elements (the star) to achieve a sound organizational design strategy: strategy, structure, processes, rewards and people. See www.privatebankinginnovation.com for details.

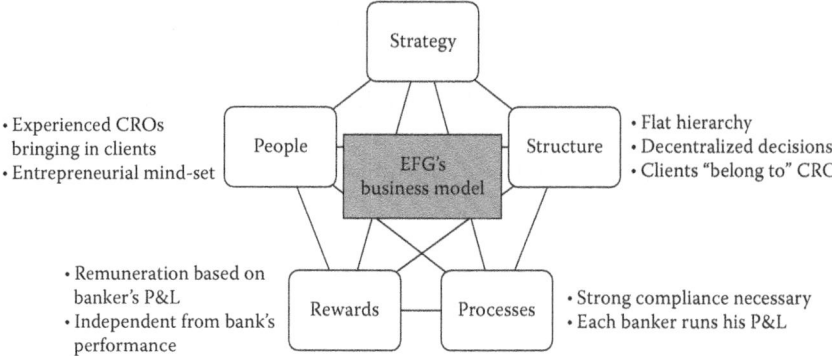

• Build a bank for and around client relationship officers (CROs)
• Growth through CRO's entrepreneurship & small acquisitions

FIGURE 3.2
EFG's Organizational Alignment Cycle.

It would be interesting for any organization to take a deeper look at the state of alignment between their business model, strategy, processes, structures, people, rewards, and implementation. In some cases the findings might prove fairly sobering.

There are a number of tools that are used to bring about Organizational Alignment. Some of the key tools are listed here:

- Business Process Improvement (BPI)
- Flowcharting
- Knowledge Mapping
- Area Activity Analysis (AAA) (see Chapter 9 for more information)
- Organizational Change Management (OCM) (see Chapter 9 for more information)

It is important to note that the objectives and performance goals are usually divided into major functions like marketing, sales, development, production, information technology, quality, and others. Each of these functions will have included in the Strategic Plan a breakdown over the five-year period of the strategies that they will be using in support of the objectives and performance goals part of the plan. Typically, the first and second years will have a more detailed, comprehensive breakdown of strategies, with Year 3, Year 4, and Year 5 defining future potential strategies.

Each year during the organizational planning cycle, the past year is dropped off and an additional year is added. In addition, each of the five years is readdressed to add more "meat to the bones" and to reflect the success or lack of success the organization had in implementing last year's Strategic Plan.

Before you start to do any structured alignment activities, set some measurements that you will use to measure and gauge the effectiveness and success of the activity. Some typical measurements would be the following:

Customer satisfaction
Employee satisfaction
Communication index
Operating cost
Better implementation of the Strategic Plan
Improvement of output quality
Increase in market share

Some of the realignment principles that might be used on a typical project are as follows:

Reduce bureaucracy.
Reduce bottlenecks.
Engage in one-stop shopping.
Empower people to make decisions at the lowest possible level.
Decentralize.
Outsource noncore processes.
Reduce layers of management.
Do not be influenced by policies.
Change management.

The ability to perform critical processes effectively directly affects the company's overall performance. It is crucial, therefore, to have a framework for understanding the bottlenecks and stress points in these processes.

—John Nisbet, founder and CEO, Kilcreggan Enterprises

ORGANIZATIONAL ALIGNMENT AND ORGANIZATIONAL PLANNING

The major purpose of Organizational Alignment is to design the organization in a way that it is best prepared to successfully fulfill the requirements and objectives of the organization's plan. The organization's plan is made up of two different but complementary plans: the Strategic Plan and the operating plan.

The Strategic Plan is designed to set the direction and expectation for the organization, and the operating plan defines what the organization will be doing to fulfill the requirements and objectives defined in the Strategic Plan. Organizational Alignment is usually used in conjunction with developing or updating the organization's Strategic Plan.

> Everyone can be working hard, but if each member of the band is playing his/her own favorite song, the result is just a disturbing noise.
>
> —HJH

Approach

If Organizational Alignment is at the core of a revolution in the rules of business, then the question becomes: what is the effect? In short, the answer is structural transformation. The results are a growing pace of application innovation, coupled with new distribution channels and competitive dynamics that are confusing and challenging even for the brightest and best managers. As technology invades and enters everything we do, organizational transformation is becoming harder to manage, simply because the issues of alignment play out on a much grander scale. Accordingly, value is found not in tangible assets such as products, but in intangibles: branding, customer relationship, supplier integration, and the aggregation of key information assets. This observation connects us to the second rule of e-business alignment.

The ability to streamline the structure and to influence and control the flow of information is dramatically more powerful and more cost-effective than moving and manufacturing physical products and services. This rule is the core driver of the structural alignment aspect of organizational transformation. However, our research shows that very few organizations have developed the necessary information-centric business designs to deal with the issues of Organizational Alignment, change, and innovation. Changing the alignment and flow of information requires companies to change not only their product mix but, perhaps more important, the planning system in which they compete. Unless an organization develops an explicit organizational plan to accommodate the accelerated flow of information, it will find itself scrambling, working harder and faster just to stay afloat. There is always hope that some magical silver bullet will appear and pierce the walls blocking the smooth flow of information, but that isn't likely as history has shown over and over again.

Outcomes

Why do successful organizations fail? The marketplace is vicious and cruel to organizations that don't align and adapt to change. History shows that organizations best positioned to seize the future rarely do so. As Alvin Toffler pointed out in *Future Shock,* either we do not respond at all or we do not respond quickly enough or effectively enough to the change occurring around us. He called this paralysis in the face of demanding change "future shock."* Too often, senior managers fail to anticipate the changing market, become overconfident, lack the ability to implement change, or fail to manage change successfully. For example, in the 1980s, IBM and Digital Equipment (DEC) were positioned to own the personal computer (PC) market, but they did nothing when upstarts such as Compaq, Dell, and Gateway took the market by storm. Why didn't they react? Because their commitment and attention were directed somewhere else.

Even as late as the early 1990s, DEC's official line was that PCs represented a niche market with only limited growth potential. DEC dug itself into a hole from which it was not possible to escape and ultimately was acquired by Compaq, a company it could have bought many times over in the 1980s. In hindsight, DEC's management team should have transformed its business design to rely less on mainframe computers and more on tapping into the PC, client/server, and Web revolution.†

As these cases illustrate, perhaps the greatest threat organizations face today is adjusting to nonstop change through Organizational Alignment in order to sustain growth. Constant change means organizations must manufacture a healthy discomfort with the status quo, develop the ability to detect emerging trends faster than the competition, make rapid decisions, and be agile enough to create and align

---

* Many credible business prophets, notably Peter Drucker (in *Managing in Turbulent Times*) and Alvin Toffler (in *Future Shock*) among others, have been anticipating this business environment of the ever-increasing rate of change for decades. The implications are that no organization, no manager, and no person should be caught off guard.

† For more information and examples of market leaders who responded and those who did not, see Gary Hamel and C. K. Prahalad's seminal work, Competing for the Future (Boston: Harvard Business School Press, 1997, 2004).

new business models.* In other words, to thrive, organizations will need to exist in a state of perpetual re-alignment and transformation, continuously creating fundamental change. Throw in the resulting time-to-market pressures, and you have some serious Organizational Alignment challenges indeed.†

ALIGNMENT TOOL #5: ORGANIZATIONAL PLANNING ANALYSIS

We find the following analysis format very helpful in setting the stage for the Organizational Alignment cycle.

Instructions: Please read each statement and indicate the extent to which it describes the reality in your organization as a whole. Your responses should reflect what you have experienced as well as what you have generally observed.

Evaluate each statement using a 10-point scale. The left side of the scale indicates that you *totally disagree* and the right side that you *strongly agree*. If you do not know, indicate so next to the question.

1. Your organization has a Strategic Plan that is up to date.
2. Everyone in the organization knows the key points in the Strategic Plan.
3. The Strategic Plan is based upon a list of assumptions that reflect the changes in the environment, customers, competition, and technologies over the coming 5 to 10 years.
4. The organization has well-defined organizational objectives for the next 5 to 10 years and these are communicated to the employees.
5. The Strategic Plan is funded.

---

* This perspective is derived from many different antecedents. The theory is closely related to the basic organizational development literature (e.g., work by Richard Beckhard, *Organization Development* [Reading, MA: Addison-Wesley, 1969]; and Michael Beer, *Organization Change and Development* [Glenview, IL: Scott Foresman, 1980]). It is also closely related to the work coming out of the University of Michigan (see Rensis Likert, *The Human Organization* [New York: McGraw-Hill, 1967] and, more recently, Denison's *Corporate Culture and Organizational Effectiveness*).

† (a) Corporate culture and organizational effectiveness. Wiley Series on Organizational Assessment and Change. Denison, Daniel R.Oxford, England: John Wiley & Sons (1990). xvii, 267 pp. Abstract. (b) Managing in Turbulent Times, Peter F. Drucker, Peter Ferdinand Drucker; HarperCollins, NY (1993). *Business & Economics*, 256 pages. This book, the author explains, "is concerned with action rather than understanding, with decisions rather than analysis." It deals with the strategies needed to transform rapid changes into opportunities, to turn the threat of change into productive and profitable action that contributes positively to our society, the economy, and the individual. (c) *Future Shock*, Alvin Toffler, Random House, NY (1970).

6. The individual work unit measurements are based upon the Strategic Plan.
7. The individual work unit budgets are related to the Strategic Plan.
8. There is an implementation plan to support the Strategic Plan.
9. The impact of the last Strategic Plan is measured and evaluated.
10. The current Strategic Plan corrects the problem that the last plan missed.

Totals: Add up the scores for each of the 10 questions:_____

Note: Grand Total of 50 or less indicates very serious problems and 80 to 100 indicates no serious issues, while 50 to 80 indicates need for improvement.

ORGANIZATIONAL ALIGNMENT AND ORGANIZATIONAL CHANGE MANAGEMENT

Change management approaches may sound like common sense but, too often, common sense is not commonly practiced.

—HJH

Only the simplest Organizational Alignment project will be successful without being accompanied by a very effective organizational change management process. A typical Organizational Alignment project will impact your processes, procedures, staff, organizational structure, and the culture of the organization, putting a great deal of stress on your people. This impact will generate a lot of resistance to the project.

To offset the many problems that occur if the affected employees are not made part of the project before it is implemented, a seven-phase change management methodology has been developed that starts as soon as the project team is assigned (see Figure 3.3).

The following gives more details related to each of the seven phases.

- Phase I. Clarify the project.
  In Phase I the scope of the project and the level of commitment by management and the affected employees required for the project to succeed are defined.
- Phase II. Announce the project.

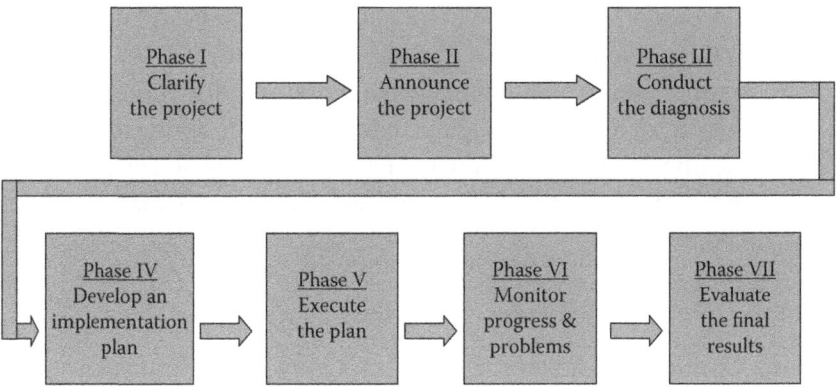

FIGURE 3.3
The seven phases of the change management methodology.

In Phase II a tailored change management plan is developed and communicated to all affected constituents. Preplanning and sensitivity to the unique needs of various groups will minimize disruption and set the stage for acceptance of the need for the change.

- Phase III. Conduct the diagnosis.
  During Phase III surveys and other types of analysis tools are used (e.g., Landscape Survey) to determine what implementation barriers exist that could jeopardize the success of the change. These diagnostic data, coupled with the rich dialogue that occurs during Phase II, provide the basis for developing an effective implementation plan.
- Phase IV. Develop an implementation plan.
  The implementation plan defines the activities required to successfully implement the project on time, within budget, and at an acceptable quality level. Typical things that will be addressed in this plan are as follows:
  - Implications of status quo
  - Implications of desired future state
  - Description of the change
  - Outcome measures
  - Burning platform criteria
  - Comprehensive or select application of implementation architecture
  - Disruption to the organization
  - Barriers to implementation

- Primary sponsors, change agents, change targets, and advocates
- Tailoring of announcement for each constituency
- Approach to pain management strategies
- Actions to disconfirm status quo
- Alignment of rhetoric and consequence management structure
- Management of transition state
- Level of commitment needed from which people
- Alignment of project and culture
- Strategies to improve synergy
- Training for key people
- Tactical action activities
- Major activities
- Sequence of events
- Phase V. Execute the plan.
  The goal of Phase V is to fully achieve the human and technical objectives of the change project on time and within schedule. It is designed to achieve these objectives by reducing resistance and increasing commitment to the project.
- Phase VI. Monitor progress and problems.
  The goal of Phase VI is to keep project implementation on track by consistently monitoring results against plan.
- Phase VII. Evaluate the final results.
  The intent of Phase VII is to provide a systematic and objective collection of data to determine if the tangible and intangible objectives of the project have been achieved and to provide insight into lessons learned and potential problem areas that may arise in future change projects.

People who welcome change make progress.
People who fight it make excuses.

—HJH

During Phase VII, management should show its appreciation to the Natural Work Teams (NWT) and individuals who expended exceptional effort during the Area Activity Analysis (AAA) project or who implemented major improvements.

4

Phase I. Strategic Planning

More than 90% of effectively formulated strategies don't get successfully implemented.

—Advait Kurlekar, Director, Cedar Consulting

The nice thing about not having a Strategic Plan is that you can't be off course if you do not know where you are going.

—HJH

The start of the Organizational Alignment Cycle is Strategic Planning (see Figure 4.1).

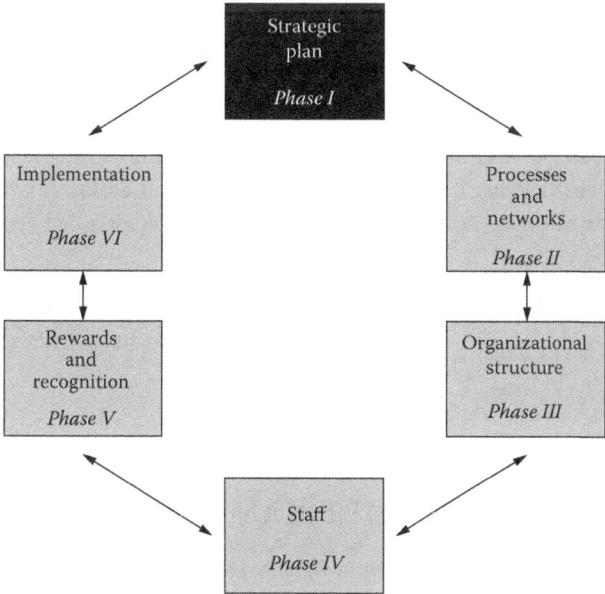

FIGURE 4.1
Phase I of the Organizational Alignment Cycle.

INTRODUCTION TO STRATEGIC PLANNING PHASE

Let's start out by answering two questions. Rate these using a scale of 1 to 10 with 1 being extremely low and 10 being extremely high.

1. How important is Strategic Planning to the organization? _____
2. Based upon actual performance, how effective have the Strategic Planning activities been to date? _____

If your organization is like most organizations, there is a big gap between the answers to questions 1 and 2. One major objective of this book is to help you close that gap.

Each time there is a major change to the Strategic Plan (or at a minimum of once every 5 years) the organization should take a serious look at realigning the organization. It is important to point out that this book was written for organizations that already had developed a Strategic Plan and who wanted to have their total organization aligned with the Strategic Plan, thereby minimizing the risk of it not being effectively implemented. As a result, we will only hit on some of the key points in Strategic Planning without going into the detail. For more detail we suggest you read *The Organizational Master Plan* by H. J. Harrington and Frank Voehl (CRC/ Productivity Press, 2012).

> Good Strategic Planning can only take place after strategic thinking has happened. What's more, strategic thinking and action take place in the throes of operational activity. This combination is called "emergent" strategy.
>
> **—Iraj Tavakoli and Judith Lawton, "Strategic Thinking and Knowledge Management"**

ORGANIZATION'S MASTER PLAN (OMP)

> Definition: The *Organizational Master Plan (OMP)* is the combination and alignment of an organization's Business Plan, Strategic Business Plan, Strategic Improvement Plan, Strategic Plan, and Annual Operating Plan.

We think of Strategic Planning as part of the Organizational Master Plan rather than a stand-alone plan. The Organizational Master Plan combines the five major management plans together into a homogeneous, agreed-to, focused approach to managing the organization's culture, business, and operations. These major management plans are as follows:

- The Business Plan (BP)
- The Strategic Business Plan (SBP)
- The Strategic Improvement Plan (SIP)
- The Strategic Plan (SP), which is the prioritized integration of the Strategic Business Plan (SBP) and the Strategic Improvement Plan (SIP).
- Annual Operating Plan (AOP)

The Organizational Master Plan must combine all of the major planning processes together and, at the same time, ensure they are customer focused. All of the plans must be based upon understanding the customer requirements and preferences, plus having an excellent understanding of what the competition is doing currently and in the future. Figure 4.2 is a model of an Organizational Master Plan, indicating the five different parts of this plan: the Business Plan, the Strategic Business Plan, the Strategic Improvement Plan, the Strategic Plan, and the Annual Operating Plan. (The Strategic Plan is the prioritized combination of the Strategic Business Plan and the Strategic Improvement Plan and in the model is represented by the combination of the two black diamonds.)

The business plan's primary objective is to inform external investors. It is prepared early in the organization's development cycle and is often used to provide potential investors information related to the organization and its management. Information from it feeds directly into the Strategic Plan. While the Strategic Plan focuses on the future of the organization, the focus of the Annual Operating Plan is short range and addresses the current operations. You will also note that the Strategic Business Plan and the Annual Operating Plan define the direction that the organization wants to take and complement each other. On the other hand, the Strategic Improvement Plan focuses on the organization's operations (how things get done) and its culture.

> Definition: The *Business Plan (BP)* is a formal statement of a set of business goals, the reasons why they are believed attainable, and the plan for reaching those goals. It also contains background information about the organization or team attempting to reach those goals.

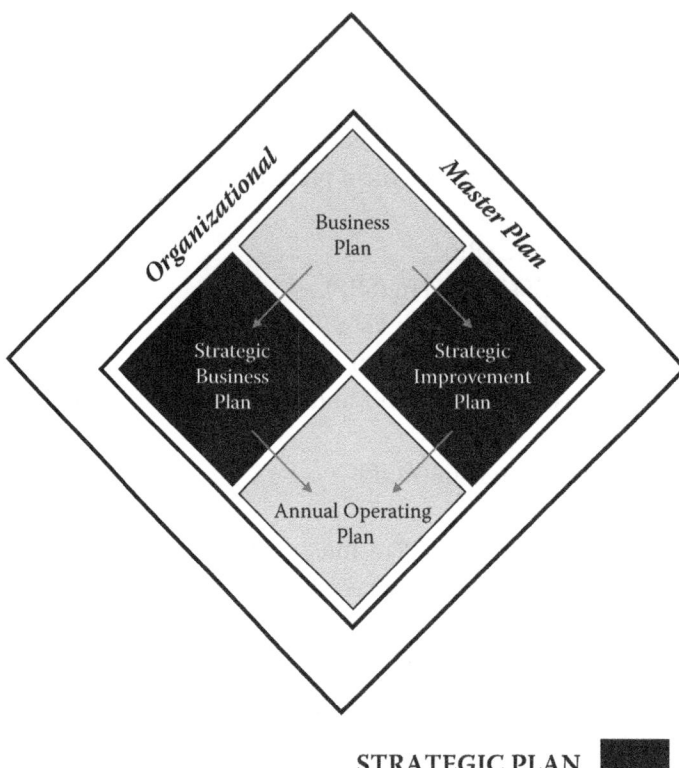

STRATEGIC PLAN

FIGURE 4.2
Five parts of the Organizational Master Plan.

Definition: The *Strategic Business Plan (SBP)* focuses on what the organization is going to do to grow its market. It is designed to answer the questions: What do we do? How can we beat or avoid the competition? It is directed at the product and/or services that the organization provides as viewed by the outside world.

Definition: The *Strategic Improvement Plan (SIP)* focuses on how to change the culture of the organization. It is designed to answer the following questions: How do we excel? How can we increase value to all the stakeholders? It addresses how the controllable factors within the organization can be changed to improve the organization's reputation and performance.

Definition: The *Strategic Plan (SP)* is a document that is the result of Strategic Planning. It defines the organization's strategy and/or direction and makes decisions on the allocation of resources in pursuit of the organization's strategy, including its capital and people. It focuses on the future of the organization and is the combination of the Strategic Business Plan and the Strategic Improvement Plan, with each item prioritized to maximize the organization's performance.

Definition: The *Annual Operating Plan (AOP)* is a formal statement of business short-range goals, the reasons they are believed to be attainable, the plans for reaching these goals, and the funding approved for each part of the organization (budget). It includes the implementation plan for the coming years (Years 1–3) of the Strategic Plan. It may also contain background information about the organization or teams attempting to reach these goals. One of the end results is a performance plan for each manager and employee who will be implementing the plan over the coming year. The Annual Operating Plan is often just referred to as the Operating Plan (OP).

Definition: *Strategic Management* is the process of specifying the organization's mission, vision, and objectives; developing policies and plans, often in terms of projects and programs that are designed to achieve these objectives; and then allocating resources to implementation of policies and plans, projects, and programs.

The key is to balance between the two increasing efficiencies today and allocating resources that will increase capacity in order to produce in the future.

—Marc Johnstone, founder, Shirlaws Global

The Strategic Plan provides a roadmap to how the following four questions will be addressed:

1. What do we do?
2. How can we beat or avoid competition?
3. How do we excel?
4. What kind of culture do we want?

It typically applies to a 5-year period, but often it is extended 10 to 20 years. The Strategic Plan defines where the organization wants to go and how it will get there.

To be successful, all organizations need to have an effective approach that sets the course for today's decisions, as well as future decisions, and to prioritize its activities. It should be made up of three parts (see Figure 4.3).

1. Setting directions
2. Defining expectations
3. Defining actions that need to be taken

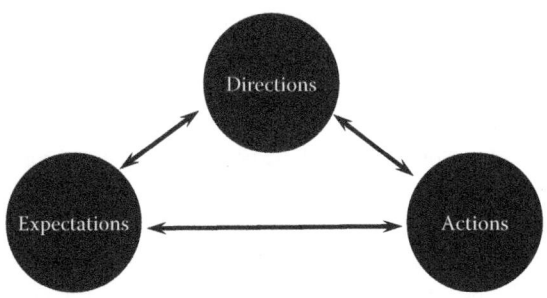

FIGURE 4.3
The three purposes of an Organizational Master Plan.

	Direction	Expectations	Actions
Business Plan	X	X	X
Strategic Plan	X	X	
- Strategic Business Plan	X	X	
- Strategic Improvement Plan	X	X	
Annual Operating Plan		X	X

FIGURE 4.4
The different types of plans and their major impact on the three purposes of the planning cycle.

Each of these purposes interrelates and reacts with each other, and each drives a number of outputs that communicates its intent to the stakeholders. Figure 4.4 shows the three major purpose areas that each of the types of plans is designed to impact.

An Organizational Master Plan is a communication, planning, and business system that reaches and involves every employee from the board to the boiler room in support of common goals and objectives. It is a three-way interactive process that provides direction, expectations, and funding. It also defines activities required to meet the agreed-upon expectations. An Organizational Master Plan includes the following 13 outputs (see Figure 4.5). To have a good Organizational Master Plan, you must understand and do an excellent job of creating and communicating these planning outputs.

SETTING DIRECTION

By failing to prepare, you are preparing to fail.

—Benjamin Franklin

Purpose	Outputs	Frequency of Updates
Setting Direction	Mission	Open
	Long-Range Vision Statement	10–25 years
	Short-Range Vision Statement	5–10 years
	KBDs' Vision Statement	5 years
	Value Statements	5 years
	Strategic Focus	3 years
	Critical Success Factors	3 years
Defining Expectations (Measurements)	Organizational Objectives	5–10 years
	Performance Goals	1–5 years
Defining Actions	Strategies	1–5 years
	Tactics	1–3 years
	Budgets	1–3 years
	Personnel Performance Plans	3–12 months

FIGURE 4.5
Organizational planning elements and frequency of changing them.

The principal role of top management is to set the direction for the organization. This can best be accomplished and communicated to the stakeholders through the Organizational Master Plan. The outputs that are used to provide this direction are defined in this section.

Definition: *Mission* is the stated reason for the existence of the organization. It is usually prepared by the chief executive officer and seldom changes, normally only when the organization decides to pursue a completely new market.

Example: An example of a "to be" mission from Boeing: "Our long-range mission is to be the number one aerospace company in the world, and among the premier industrial firms, as measured by quality, profitability and growth."

Example: An example of a "to do" mission from McDonald's: "To satisfy the world's appetite for good food, well-served, at a price people can afford."

Definition: *Key Business Drivers' (KBD) Vision Statements* (also called *Controllable Factors' Vision Statements*) are statements of how the Key Business Drivers will be operating 5 years in the future. They are developed during the Strategic Improvement Planning process. There are usually 8 to 12 of these vision statements.

Example of a Processes Vision Statement: Our critical processes provide a competitive advantage that employees and customers recognize as world class. Continuous improvement using statistical data is a way of life.

Processes are defined with assigned owners, evaluated against established measures, and operate in the real world.

> Definition: *Key Business Drivers (KBD)* (also called *Controllable Factors*) are things within the organization that management can change that control and/or influence the organization's culture and/or the way the organization operates.

Typical KBDs that impact organizational performance are:

1. Measurement system*
2. Training*
3. Management and leadership methods*
4. External customer partnership interface*
5. Supplier partnership*
6. Business processes*
7. Production processes
8. Corporate interface to the divisions
9. Employee partnership
10. Research and development activities
11. After-sales service processes
12. Knowledge management*

Note: KBDs with an asterisk have predefined 12 level maturity grids already developed for them.

> Definition: *Value Statements* are documented directives that set behavioral patterns for all employees. They are the basic beliefs that the organization is founded upon, the principles that make up the organization's culture. They are deeply engrained operating rules or guiding principles of an organization that should not be compromised. They are rarely changed. Value statements are sometimes called operating principles, guiding principles, basic beliefs, or operating rules.

Example: Owens Corning Fiberglass uses "guiding principles" in place of "values." Their guiding principles are the following:

- Customers are the focus of everything we do.
- People are the source of our competitive strength.
- Involvement and teamwork is our method of operation.
- Continuous improvement is essential to our success.
- Open, two-way communication is essential to the improvement process and our mission.

- Suppliers are team members.
- Profitability is the ultimate measure of our efficiency in serving our customers' needs.

> Definition: *Strategic Focus* comprises the key things that will set the organization apart from its competitors over the next 3 years. This list is defined by top and middle-level managers. It is directed at the organization's core competencies and capabilities.
>
> Definition: *Critical Success Factors* are the key things that the organization must do exceptionally well to overcome today's problems and the roadblocks to meeting the vision statements.
>
> Definition: *Core Competencies* are the technologies and production skills that underlie an organization's products or services (e.g., Sony's skill at miniaturization).
>
> Definition: *Core Capabilities* are the business processes that visibly provide value to the customer (e.g., Honda's dealer management processes).
>
> Definition: *Strategic Excellence Positions* are the unique and distinctive capabilities that are valued by the customer and provide a basis for competitive advantage (e.g., Avon's distribution system).

DEFINING EXPECTATIONS (MEASUREMENTS)

One of the major purposes of an Organization's Master Plan is to define what management and the stockholders expect from the organization's performance over the next 5 to 10 years and then to communicate how success will be measured. The outputs that are used to communicate these expectations are organizational objectives and performance goals.

> Definition: *Organizational Objectives* are used to define what the organization wishes to accomplish over the next 5 to 10 years.

Example: IBM released the following objectives that it planned to accomplish during a 10-year period.

- To grow with the industry.
- To exhibit product leadership across our entire product line. To excel in technology, value, and quality.
- To be the most effective in everything we do. To be the low-cost producer, the low-cost seller, the low-cost administrator.
- To sustain our profitability, which funds our growth.

> Definition: *Performance Goals* are used to quantify the results that will
> be obtained if the organizational objectives are met.

Example: On February 25, 1986, then President of the United States Ronald Reagan released Executive Order 12552, stating: "There is hereby established a government-wide program to improve the quality, timeliness, and efficiency of services provided by the federal government. The goal of the program shall be to improve the quality and timeliness of service to the public, and to achieve a 20% productivity increase in appropriate functions by 1992. Each executive department and agency will be responsible for contributing to the achievement of this goal."

DEFINING ACTIONS

Another purpose of an Organizational Master Plan is to drive the organization's change process to define the actions that should be taken to implement the plan over the next 5 years. It is designed to focus the resources of the organization in line with its expectations. The outputs that are used to communicate these actions are strategies, tactics, budgets, and personnel performance plans.

> Definition: *Strategies* define the approaches that will be used to meet the
> performance goals.
> Definition: *Tactics* define how the strategies will be implemented. They
> explain how the strategies will be accomplished.
> Definition: *Budgets* provide the resources required to implement the
> tactics.
> Definition: *Personnel Performance Plans* are contracts between management and the employees that define the employees' roles in accomplishing the tactics and the budget limitations that the employees have placed upon them.

APPROACH TO DEVELOPING A BUSINESS PLAN

> Definition: *Business Plan* is a formal statement of a set of business goals,
> the reasons why they are believed attainable, and the plan for reaching those goals. It also contains background information about the organization or team attempting to reach those goals.

Business Plans may be internally or externally focused, but normally they are externally focused and target goals that are important to external stakeholders, particularly financial stakeholders. They typically have detailed information about the organization or team attempting to reach the goals. They are most often used by startup organizations to inform potential investors or lending organizations.

Often the first exposure to a Business Plan is for the purpose of convincing investors and lenders that you have a viable idea that they should invest in. In reality this is usually the main reason that most entrepreneurs develop a Business Plan. But the Business Plan serves a much greater purpose. It forces the entrepreneur to step back and take a look at what he or she is doing from a business perspective. Think of a start-up's promotional plan as concept driven; it's more general in nature. The presentation leaves many questions of practical execution unanswered. These plans are fine for their purpose. However, most aren't intended as a blueprint for running the organization. But these initial Business Plans will mature into a Strategic Plan as the organization grows and develops.

Often the hardest part of initiating or updating the Business Plan is honestly determining an accurate assessment of your current position today. It's not always as obvious as we may think.

Outline of a Typical Business Plan

The following is a guideline to the headings that make up a typical Business Plan.

- **Cover Letter**
  - Basic information about the organization—address, date of the document, who approved it, who prepared it, etc.
- **Executive Summary**
  - Organization's purpose and objectives
  - Products and/or services description
  - Competitive advantages of the products and/or services the organization will be providing
  - Present status of product and/or services
  - Present management team backgrounds and why they bring value to the organization
  - Key success factors

- Income projections by year
- Financial investment to date
- Key persons' present financial investment in the organization
- Financial status and future requirements
- Present financial needs
- Profit and loss projection
- Cash flow projection
- Key milestones
- Assumptions and comments
- **Two-Year Market Analysis**
  - Target market
  - Market research
  - Market trends
  - Price impact on the market
  - Percent of the market the organization can get by year and why
  - Customer profile
  - Marketing and sales strategy and plans
  - Marketing and sales plan strengths and weaknesses
  - Public relations plan
- **Competitive Analysis**
  - Product current and future state analysis
  - Competition description
  - Competitive products/services
  - Competitive research data
  - Competitor's strengths and weaknesses
  - Projected changes in the major competitor's product
  - Comparison of organization's product and/or services to those provided by the competition
  - Risk analysis
- **Products and/or Services**
  - Product/service description
  - Positioning of products and/or services
  - Development status of products and/or services
  - Project plan
  - Work breakdown structure
  - Cost to provide the product and/or services
  - Competitive evaluation of products and/or services

- - Delivery schedule
  - Implementing strategy
  - Plans for additional products and/or services
- **Marketing and Sales**
  - Marketing strategy
  - Sales strategy
  - Advertising approach
  - Announcement plans
- **Operations**
  - Mission and visions
  - Organizational structure
  - Business model
  - Intellectual capital
  - Legal structure
  - Insurance
  - Human resources plan
  - Financial management strategy
  - Information technology plan
  - Customer support plan
  - Facilities today and in the future requirements
- Organizational SWOT analysis (strengths, weaknesses, opportunities, and threats)

As you can see, preparing the startup Business Plan requires a great deal of thinking and work on the part of the key people in the organization, but the time devoted to it is well spent, for it forces the leaders of the organization to get organized. Also the work done on it will serve as a major input to the Strategic Business Plan that will be prepared as the organization matures. For more information on preparing a Business Plan, we suggest you read *Organizational Master Plan* (CRC/Productivity Press, 2012).

APPROACH TO DEVELOPING A STRATEGIC BUSINESS PLAN

We suggest using a 10-activity approach to develop a Strategic Business Plan.

Activity 1. Define the Strategy Scope and Time Frames

Activity 2. Define the Assumptions

Activity 3. Review the Mission, Long-Term Vision, and Value Statements

Activity 4. Define the Short-Term Vision

Activity 5. Define Core Competencies and Capabilities

Activity 6. Develop a Risk Analysis

Activity 7. Define Critical Success Factors

Activity 8. Set Objectives and Goals

Activity 9. Develop Strategies

Activity 10. Develop Tactics for Each Strategy

One of the decisions that needs to be made early in the Strategic Planning cycle is whether the organization should be a product-centric or customer-centric organization, as it will be one of the primary drivers of the planning and implementation process. Some of the key differences are listed in Figure 4.6.

> Definition: *Product-Centric Organizations* are organizations that have multiple product lines that divide into separate business lines and/or models. Often there are little or no interrelationships between the product lines (e.g., a computer manufacturer that sells management consultant services). They are driven by product portfolios.
>
> Definition: *Customer-Centric Organizations* are organizations whose offerings are combined and integrated advice, services, and/or software in support of their products/services that result in customized offerings. They are driven by customer portfolios.
>
> Definition: *Customer-Focused Organizations* use extensive market research in defining their product offerings and design. They build a demand for the products they are able to produce. They invest in

PRODUCT-CENTRIC	CUSTOMER-CENTRIC
New products	Customized products and services
Lead-edge/new features	Customized solutions
Sophisticated customers	Preferred/profitable customers
Best products	Best solutions
Market-driven pricing	Value-driven pricing
Understand the competition	Understand the customer
Product teams	Customer teams
Product profit centers	Customer segments
New product development focus	Customer relationship management focus
How many ways can the product be used?	What combination of products is best?

FIGURE 4.6

Product-centric versus customer-centric organizations.

providing their front-line employees with the knowledge, products, and tools to provide effective and consistent customer service.

STRATEGIC BUSINESS PLANS VERSUS STRATEGIC IMPROVEMENT PLANS

There is a big difference between a Strategic Business Plan and a Strategic Improvement Plan. The Strategic Business Plan sets the product and service strategy for the organization: the markets that they hope to penetrate, the new products that will be introduced, the production, etc. It is a plan that directs and guides the business as it is viewed by its customers. The Strategic Business Plan is primarily directed at meeting the needs of only two of the stakeholders associated with the organization—the customer and the stockholder. It is a plan that is primarily focused on the external opportunities.

The Strategic Improvement Plan, on the other hand, is an internally focused plan that is designed to transform the environment within the organization to change its personality (behavioral characteristics) to be in line with the business plan. It takes into consideration the needs of all of the organization's stakeholders from an improvement standpoint. The Strategic Improvement Plan defines the transformation in the business personality of the organization. It provides an orderly passage from one state or condition to another. The Strategic Improvement Plan supports the Strategic Business Plan so the two of them, although different in content and intent, must be kept in close harmony.

DEVELOPING A STRATEGIC IMPROVEMENT PLAN

The following are the five activities that make up the Strategic Improvement Planning Cycle:

Activity 1. Assessment
Activity 2. Vision Statements
Activity 3. Performance Goals
Activity 4. Desired Behavior
Activity 5. Three-/Five-Year KBD Plans

In the landscape of differentiation, behavior is the final frontier.
—**Terry Bacon**, Founder of Lore International Institute

Activity 1. Assessing the Organization

During the assessment activity, the KBDs are defined, the present culture is evaluated, and an as-is view of each of the KBDs is developed.

Activity 2. Developing Vision Statements for Each of the KBDs

During this activity the Executive Team develops a statement of how they would like the organization to be functioning related to each of the KBDs 5 years in the future. These first draft vision statements are then presented to the organization's stakeholders for their comments and suggestions. Based upon these inputs, the set of 8 to 12 vision statements are revised and finalized.

Activity 3. Developing a Set of Performance Goals

In this activity the Executive Team develops a set of performance goals by year for the next 3 to 5 years. The yearly goals are used to measure progress.

Activity 4. Defining Desired Behaviors

Improvement in culture and performance requires changes in the behavioral patterns of the Executive Team and its employees. Because the team's behavioral patterns need to change before the desired results can be measured, a set of behavioral patterns that reflect the required behavior changes is identified.

Activity 5. Preparing 3- to 5-Year Improvement Plans for Each of the Vision Statements

During this activity a Performance Improvement Team (PIT) will address each of the vision statements and define the problems related to

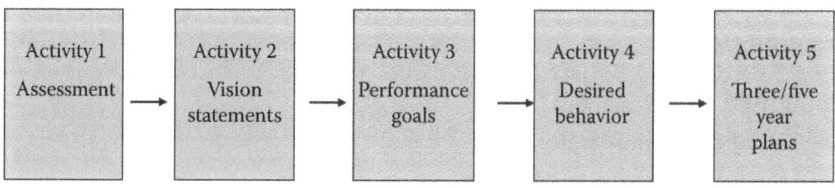

FIGURE 4.7
The five activities of the Strategic Improvement Planning Cycle.

the vision statement that the organization is facing today. Then they will make a list of the roadblocks that the organization will face in making the transformation. Once these are defined, the root causes for each of the problems and roadblocks will be identified. For each root cause, the PIT will select a tool/methodology to overcome the obstacle. A list of over 1100 tools/methodologies can be found in Appendix B of this book. The PIT will also define the time required to implement the selected tools/methodologies.

CREATING THE STRATEGIC PLAN

We have briefly covered how to generate the two halves of the Strategic Plan: the Strategic Business Plan and the Strategic Improvement Plan. But do we have two halves that make a whole? Are there voids, overlaps, or both, in the two plans? If we work with them as two separate entities, there will be a lot of competition for people and other resources. Which plan or part of the plan has priority? Is there any interdependencies between the two plans? Which plan gets credit for the organization's success and which one gets the blame if things don't work out well? In reality, you should not have two separate plans; they need to be combined so the total organization is singing the same song.

There are a number of things that should be considered when the individual plans are combined, not the least of which is the performance improvement goals that were developed earlier. The Executive Team should review the individual plans to define which activities impact each of the performance improvement measurements and schedule the activities so that the performance goals are met. Often we will find that there are multiple strategies and/or tools defined to address a single item. Look at these very carefully to understand the impact that each of the strategies and/or tools will have on the item. Often you will find out that you can get the desired results without using all the strategies and/or tools that were directed at them. When you have multiple strategies and/or tools directed at the same item, install them far enough apart so that the impact of each is measured before the next one is implemented because there may be no need for implementing it.

As a way to help you prioritize the activities, we suggest that you make a matrix of seven columns. In the left-hand column list all of the strategies, tools, and methodologies defined in the two plans (see Figure 4.8). The headings of the other columns are as follows:

Column 2 – Employee Impact
Column 3 – Dollars Saved
Column 4 – Business Impact
Column 5 – Customer Impact
Column 6 – Total
Column 7 – Suggested Year to Implement

For every block in the chart, rate it from 1 to 5.

1 = Very low impact
2 = Low impact
3 = Average impact
4 = High impact
5 = Very high impact

In this case the higher the total, the bigger the impact on the organization. In most cases the high-impact items get scheduled early in the cycle but not in all cases. In some cases an item cannot be implemented until other conditions have been completed first.

Using the matrix as a guide, a team of knowledgeable people can develop a 5-year strategic action plan that defines by quarter when each strategy,

Strategies, Tools, and Methodologies	Employee Impact	Dollars Saved	Business Impact	Customer Impact	Total	Improvement Year
Lean	3	2	3	1	9	1
Knowledge Management System	4	2	4	2	12	3
Operating Manual	3	2	2	1	8	2
Team Training	4	2	3	1	10	1
Self-Management Teams	5	2	1	0	8	4
Organizational Alignment	4	3	5	2	14	1
Supply Chain Management	1	3	4	2	10	2

FIGURE 4.8
Prioritizing Strategic Plans.

tool, or methodology should be started and completed. Both dates are important.

ANNUAL BUSINESS PLAN

We will not cover how to develop an Annual Business Plan, because these plans usually do not drive an Organizational Alignment activity. For more information on Annual Business Plans, see the book *The Organization's Master Plan,* by H. James Harrington and Frank Voehl (published by Paton Press).

SUMMARY

We have just presented a short description of the Strategic Planning cycle. It is a complex cycle that requires a lot of creativity, thought, and dedication to be effective. Although the Strategic Plan is reviewed and updated each year to reflect last year's activities and any new developments, it usually doesn't go through a major revision more often than every 2 to 5 years. If you make major revisions more frequently, it is difficult to hold people accountable to meet the goals and to make the required behavioral changes. Each time there is a major change to the Strategic Plan, it should trigger the start of the Organizational Alignment cycle. Assuming you did make a major change in your Strategic Plan, you should now be ready to move on to Phase II: Process and Networks Design in the Organization Alignment Cycle.

Planning is fun—it is implementing the plan that brings out the sweat.

—HJH

5

Phase II. Processes and Networks

When the processes work, so does the organization.

—HJH

Phase II of the Organizational Alignment Cycle is defining/designing the organization's processes and networks (Figure 5.1).

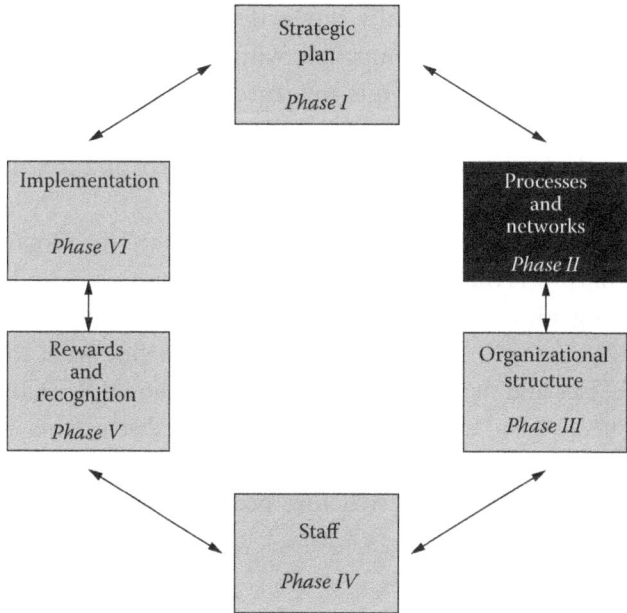

FIGURE 5.1
Phase II of the Organizational Alignment Cycle.

INTRODUCTION TO PROCESS AND NETWORKS PHASE

Why is it difficult for established organizations to see the writing on the wall? The main reason is because they want to "stick to the knitting," that is, to continue to do what made them successful. They don't want to cannibalize existing product lines, and they tend to fall back on simple formulas: lower costs, increase operational efficiency, and increase product variety. They should look at technology as a way to make their lives easier and give them more value for their money. Established organizations must challenge traditional definitions of value. They must learn to take advantage of new technologies to create Organizational Alignment and deliver new streams of value. And they must use a process approach to help with this transition.

The process approach to Organizational Alignment calls for the integration and linking of the vision/strategy, the social system/culture, processes, networks, people, management systems, and rewards to accomplish the best results. Alignment becomes part of the DNA when strategic goals and cultural values are mutually supportive and where each part of the organization is linked and compatible with the other parts. In an effective organization there is a definite congruence between purpose, strategy, processes, networks, structure, culture, and people. It is the paradox and challenge to the leaders to orchestrate this alignment and to still promote innovation and change.

In the previous chapter we defined how to prepare the Organizational Master Plan. Now is the time to consider how the processes and networks within the organization will need to change to best support the Strategic Plan. To accomplish this, the team will need to define what the current major processes and networks are and how they should be changed to be the most effective and efficient in the newly aligned structure. To accomplish this, we like to use the following tools, but try not to relate them to the present organizational structure because the structure will often change during Phase III.

- Flowcharting
- Process mapping
- Value Stream Mapping
- Business Process Improvement
- Knowledge Mapping
- Information Mapping

Defining Major or Core Processes

All organizations know that their success or failure rests on how well the process and organizational structure are aligned. As a result, the final organizational structure is heavily influenced by the way the processes are designed. We feel strongly that flowcharting of the AS-IS and the SHOULD-BE processes is a key tool for gauging the effectiveness of the alignment effort and needs to be prepared and understood before the final organizational structure can be defined. To fully understand your organization, you need to take flowcharting to the next level, process mapping, which is a tool used to help management and workers gain alignment from a fresh process insight.

> Definition: *Flowcharting* is a method of graphically describing a process (existing or proposed) by simple symbols and words to display the sequence of activities in the process. (For more information on flowcharting, see Book I of *Waging the War on Waste: Basic Performance Improvement Methods*, by H. J. Harrington and K. Lomax (published by Mc-Graw Hill.)
>
> Definition: Process mapping is a hierarchical method for displaying processes that illustrates how a product or transaction is processed. It is a visual representation of the work flow either within a process or of the whole operation. Process mapping comprises a stream of activities that transforms a well-defined input or set of inputs into a predefined set of outputs. It is a flowchart with inputs and outputs added to each activity, thereby increasing its value in refining processes.

When applied to the core business processes, flow charting may help identify areas where breakthrough is possible in order to gauge the true effectiveness of the effort. Traditional thinking about processes comes from "process blindness." Because managers have not had to pay much attention to processes, this blindness has plagued most organizations for many years. On an operational level, most senior managers have no idea how their organizations operate. In other words, real day-to-day operational performance is no longer understood, nor is it controlled on a real-time basis.

Managers have traditionally focused on strategy, planning, organization, and market/product innovation. The process movement has now forced these managers to take a fresh look at their organization's business processes in order to be accountable for results. One of the greatest practical hurdles to developing process insight is the business mind-set that operates in terms of functions. Also, if competitors do something differently and help create new Critical Success Factors (CFSs), the process may need to be changed to keep pace and break away from the old views. The problem becomes gauging accurately the impact of these efforts.

To break away from the traditional view where single functions dominate as a natural way of thinking, Keen and Knapp pose several questions for management to answer in order to gauge the effectiveness of alignment efforts.*

- Who exactly is the customer or person to which the outcome matters?
- What must happen for the customer's request to be completely satisfied?
- Who does the work and how does it align (come together)?
- How will the work be coordinated logistically?
- Can information technology be exploited to improve coordination? Empower the people to do the work? Augment training? Alter incentives?

The answers to these questions help us to gauge the alignment of the work flow and coordination activities. They look at process, not function or activity. They do not prejudge the solution. Process Management allows for either an evolutionary approach of continuous improvement or a revolutionary one, such as Process Redesign. Also, some processes, such as logistics, cannot be redesigned quickly, and thus there are barriers to meeting customers' needs. By using evolutionary methods, such as adaptive engineering, there is a possibility for rapid and significant change that delivers a breakthrough that can be gauged and measured.

Much of Business Process Improvement (BPI) has traditionally focused on documenting the right way to do something. However, one of the dangers is that it is easy for a process to become repetitive, as the focus is always on "know-what" rather than "know-why." Consider taxicab drivers in the United Kingdom. Unlike cab drivers in the United States, they go through rigorous training to get their taxi licenses. They are taught not just "know-what" (which one might define as the shortest route on a map for getting someone from point A to point B), but they also develop "know-why," which might mean that when it rains it is better to take the longer route, which doesn't have the sewage backups that create traffic delays.

"Know-why" breeds innovation and adaptability, which are exactly what organizations need when they face shifts in strategic direction. Some

---

* Peter G. W. Keen and Ellen M. Knapp, *Every Manager's Guide to Business Processes* (Boston, MA: Harvard Business School Press).

process reengineering projects run the risk of throwing out the "know-why" along with the old processes. Successful organizations with strategic momentum have a healthy respect for traditional processes and innovative ones at the same time.

Process Improvement Rules

The following rules provide an excellent guideline to help you streamline your processes.

- Design the process around value-adding activities.
- Use organizational change management from Day 1.
- Pick the low-hanging fruit first.
- Use self-inspection versus appraisal.
- Don't write the same thing twice.
- Eliminate redundant tasks.
- Stop doing dumb things.
- Ask the person doing the job what should be changed.
- Shrink the hidden office (no-value-added and business-value-added activities).
- Paint a real picture of the process.
- Eliminate duplication in work and data.
- Combine similar activities.
- Reduce the amount of handling.
- Eliminate unused data.
- Clarify forms.
- Revise office layout (work flow).
- Remember all of your stakeholders may not speak English.
- Make the process less complex and demanding.
- Minimize interdependencies.
- Make it easier to learn.
- Question if it even needs to be done.
- Minimize interruptions.
- Eliminate wait time.
- Improve timing (when the employee receives the input).
- Remove distractions.
- Use the 5S (Keep the place clean neat).
- Reduce employee fatigue.
- Work is performed where it makes the most sense.
- Provide a single point of contact for customers and suppliers.

- If the inputs coming into the process naturally cluster, create a separate process for each cluster.
- Ensure a continuous flow of the "main sequence."
- When it is moving from place to place, it is waste.
- When anything is not moving through the process, it is wasting money.
- Reduce setup and changeover times.
- Reduce batch sizes.
- Substitute parallel processes for sequential processes.
- Perform process activities in their natural order.
- Reduce checks and reviews.
- Push decision making down to the lowest reasonable level.
- Build quality in to reduce inspection and rework.
- Simplify activities.
- No news is bad news.
- Start big to see if any of the subprocesses can be eliminated.
- Know your customer as well as you know your process.
- Eliminate loops, branches, and rework.
- Cost-justify bureaucracy (B).
- Cycle time is as important as processing time.
- Error-proof the process.
- Write your procedures for the user (use simple language).
- Use IT as a last resort.
- Document when, where, how, and the quality of all inputs and be sure the supplier knows what is required.
- Define what the desired future state will be.
- Set design goals and measurements.
- Pilot the new process.
- Design the organization around the process, not the latest technologies.
- Turn sacred cows into hamburgers.
- Define Critical Success Factors.
- Use benchmarking to set targets and define new ideas.
- Get fast results by working with subprocesses or activities (Fast Action Solution Teams).
- Right size, not down size.
- Consider hybrid centralized/decentralized operations.
- Bring downstream information needs upstream.

- Capture information once at the source, and share it widely. Share all relevant information.
- Involve as few people as possible in performing a process.
- Redesign the process first, then automate it.
- Ensure 100% quality at the beginning of the process.
- Increase flow and speed to identify bottlenecks.
- Eliminate bottlenecks.
- Design for manufacturability and serviceability.
- Use Design for Reliability.
- Install metrics and feedback to find and correct problems.
- Find opportunities for continuous improvement.
- Use simulation, practice, or role playing to test new process designs.
- Standardize processes.
- Use co-located or networked teams for complex issues.
- Assign a process consultant for cross-functional processes.
- Involve process workers in analyzing, designing, and implementing improvements.
- Form work cells for special cases or exceptions.
- Use multifunctional teams.
- Use multiskilled employees.
- Create generalists instead of multiple specialists.
- Employ mass customization.
- Redesign work better than reengineering, in most cases.
- Use the most advanced cost-effective equipment.
- Save human energy to reduce cost and improve quality.
- There always is a better way.
- Measure the processes' effectiveness, efficiency, and adaptability.
- Challenge all paradigms.
- It is sometimes more expensive to use IT than to do it by hand.
- Know the total cost of what changes you make.

Process Improvement Approaches

BPI reengineering is by far the most influential and most controversial intervention within the process movement. It calls for a radical change in the business process as a matter of survival for organizations and targets dysfunctional or broken, outdated processes for investment and process reengineering. The process reengineering effort usually requires a major investment in information systems and technology. It is the fundamental rethinking and radical redesign of business processes to achieve dramatic breakthrough improvements in critical, contemporary measures of performance, such as cost, quality, service, and speed. When you need to reduce cost, cycle times, or both by 60% or more, you usually need to reengineer the process.

Whereas traditional quality approaches start with stable processes and the need to improve them (or first stabilize them), BPI throws away how things are done today and starts over with a blank slate to design new processes and organizational structures that will achieve, in one fell swoop, breakthrough and competitive advantage. BPI reengineering calls for starting with a clean sheet of paper and asks the question, "If we were to start this organization today, how would we design the process?" It assumes a fundamental disconnect between an organization's current course and what is required for success in the marketplace. As such, it will be required of organizations that never got off the quality bandwagon, as well as those whose markets are becoming increasingly more turbulent and are changing faster than their current logistics programs have been able to handle.

The major Organizational Alignment opportunity lies in the cross-functional streamlining of work activities. It attacks division of labor and functional organization of work as the biggest constraint in service improvement and higher. The credo is "don't automate—obliterate!" Most organizations that attempt to implement reengineering settle for a form of business process redesign instead because of questions concerning ethics, risk, practicality, and cost.

Major Networks

There is one other consideration that needs to be addressed: it is the networks that exist in all organizations that are not part of the formal processes. Networks are in the informal systems that people set up as they learn the personality of the people and the processes that they interface with. It is things like knowing that one group reacts and another group does not react promptly to a request. It is unfortunate, but that is real life. There are a number of excellent approaches that help define the effectiveness of the different interfaces. We have found that a "supplier/customer interface analysis survey" is a very effective tool to understanding the network. Sometimes instead of paying for a new IT system to manage the organizational knowledge assets, it pays to leverage the social networks through which knowledge already flows. One large petroleum company based in the United Kingdom decided that it was not worth it to spend millions of dollars to create a huge knowledge database, which would require disrupting processes, people, and structure all at once. Instead they created a peer-to-peer action review, where they pulled people together from around the world on a quarterly basis to share solutions to critical strategic issues. By facilitating the expansion of social networks of key employees, they were able to achieve high-quality knowledge sharing without the disruption that a huge technology effort would have required.

Of course, as previously outlined, tangible rewards, such as compensation and bonuses, are a good way to get a desired shift in behavior from employees, but more often than not, organizations overlook how much "good old recognition" works.

As one executive from a large investment organization put it, "Contrary to popular belief, I don't think people resist change. We hire smart people and know that smart folks actually want to be part of building a better organization, even if it means things have to change." This same executive said that he goes out of his way to acknowledge an employee's desirable behavior in front of the employee's peers. Furthermore, he tries to acknowledge positive change as a group effort: "If you reward individuals too often it can make them less willing to share information. It creates a tendency to hoard what they know in order to set themselves apart—and that's the last thing we need around here."* So, before launching a new compensation plan, which tends to cause anxiety and stress for many employees, organizations should consider whether or not "soft" rewards, such as recognition in front of peers, will get them where they want to go.

Organizational Alignment and Knowledge Mapping

There will be an increased emphasis over the next few years on taxonomies, ontologies, and knowledge.

—French Caldwell, VP Information and Knowledge Management,
Gardner Group

---

* For further discussion, see J.N. Lowenthal, *Reengineering the Organization: A Step-by-Step Approach to Corporate Revitalization* (ASQC Press, 1994).

A knowledge map portrays a perspective of the players, sources, flows, constraints, and alignment of knowledge within an organization. It is a navigation aid to both explicit (codified) information and tacit knowledge, showing the importance and the relationships between knowledge stores and the dynamics. The final "map" can take multiple forms, from a pictorial display to a yellow pages directory, to linked topic or concept map, to inventory lists or a matrix of assets aligned with key business processes.*

> With few exceptions, most firms have had difficulty in developing a viable strategic information system. There are manifold reasons for this, but certainly a major one is the omnibus nature of the data required.
>
> **—Alberto Cameiro, Associate Professor, University of Lisbon**

Purpose of Knowledge Mapping

Knowledge mapping is an ongoing joint quest to help discover the constraints, assumptions, location, ownership, value, and use of knowledge assets, artifacts, people, and their expertise; uncover blocks to knowledge creation; and find opportunities to leverage existing knowledge. Knowledge mapping may involve developing an ontology conducting social network analysis; executing a survey; or engaging a group of people in sense-making, action research, or ethnography. The process of making the knowledge map is as important as the final product because it's impossible to create a single map that will meet the needs of every situation. Agreement is required by decision makers regarding the purpose of the knowledge mapping exercise and a map or maps created to meet those objectives.

> The essence of knowledge exchange is human dialogue. What we call knowledge encompasses everything from the precision of mathematics and physics to complex human interactions.
>
> **—Michael Utvich, *Handbook of Business Strategy*, 2005**

---

* A full treatment of this subject can be found in H. James Harrington and Frank Voehl, *Knowledge Management Excellence* (Chico, CA: Paton Press, 2007). In this work, we point out that it can be extremely difficult in some instances to quickly identify important knowledge assets because people forget about what they know until they need to know it. Consequently, it can be useful to collect stories of how people work together in order to remind others of the knowledge that they rely on. This story base provides evidence that helps the knowledge mapper know where to look and what to include in the map.

Knowledge Mapping Approach

Knowledge mapping is an alignment of data gathering, survey, exploring, discovery, conversation, disagreement, gap analysis, education and synthesis. Its major aims are to do the following:

- Track the loss and acquisition of information and knowledge, personal and group competencies and proficiencies.
- Show knowledge flows and appreciate the influence on intellectual capital due to staff loss.
- Assist with team selection and technology matching.

Contrast this with a *knowledge audit,* which tracks deviations from policy or established process, checks for compliance with standards and procedures, and seeks to measure and value knowledge assets and marketable intangibles.[*] A knowledge audit focuses on finding, itemizing, and attaching values to *knowledge assets* and checking compliance with approved processes. The key activity is determining the worth and market value of *intellectual property and capital* and spotting policy and practice deviations. Mostly this activity is concerned with portfolio management and tangible (hard) assets.

A knowledge audit covers legal and security (protection) issues, ownership, market value, portfolio dynamics and synergies, and potential for realizing capital gains and enhancing revenue streams. The audit looks at conformance and compliance, concentrates on objects that are marketable or nearly so, rather than the environment for new knowledge creation. Every audit attempts to locate, measure, and evaluate assets with some potential market value and checks for deviations from accepted processes.[†]

The Knowledge Management (KM) audit takes place after your organizational knowledge policies and processes are put in place and related practices have been established. The audit measures how faithfully the organization is following these authorized practices, while listing gaps and departures. Suggestions for revisions, controls, and reviews help to bring things back in line. KM mapping is often done at the start of a major project to collect baseline data, although we suggest that mapping is use-

---

[*] According to Wikispaces. See http://kmwiki.wikispaces.com/Knowledge+mapping.
[†] Ibid.

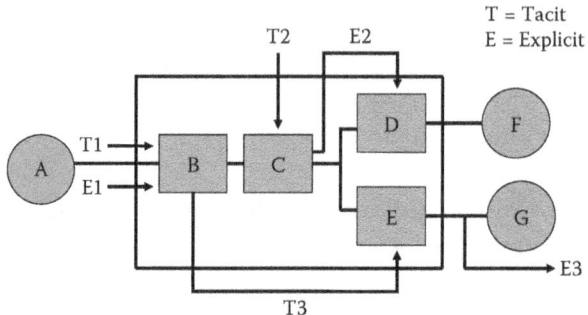

FIGURE 5.2
Activity 1. Create the knowledge map.

ful as an ongoing exercise, especially coupled with process mapping. The emphasis is on exploration, discovery, and opportunity finding.*

Knowledge audits are scheduled to value intangibles (intellectual property, social and intellectual capital) and mostly are done on an annual basis, before mergers and acquisitions and as part of "accounting" reviews or strategic due-diligence exercises. A processed-based knowledge map is a map or diagram that visually displays knowledge flow within the context of the business process. It shows how knowledge should be used within the process and the source of this knowledge. Any type of knowledge that drives the process or results from execution of the process can be mapped. This includes tacit (soft) or explicit (hard) knowledge.

There are three activities to the knowledge mapping process. They are as follows:

- Activity 1. Creating the knowledge map
- Activity 2. Analyzing the map
- Activity 3. Applying the map

Figure 5.2 is a typical knowledge map of a four-activity process. You can see how both explicit (hard) and tacit (soft) knowledge is used

---

* Ibid.

throughout the process. Knowledge mapping is used to accomplish the following:

- Find key sources, opportunities, and constraints to knowledge creation and flows.
- Encourage re-use and prevent reinvention, saving search time and acquisition costs.
- Highlight islands of expertise and suggest ways to build bridges to increase knowledge sharing and exchange.
- Discover effective and emergent communities of practice where informal learning is happening.
- Provide baseline data for measuring progress with projects and justifying expenditures.
- Reduce the burden on experts by helping staff to find critical solutions and information quickly.
- Improve customer response, decision making, and problem solving by providing access to applicable information and internal and external experts.
- Highlight opportunities for learning and leverage of knowledge through distinguishing the unique meaning of "knowledge" within that organization.
- Provide an inventory and evaluation of intellectual and intangible assets and assess competitive advantage.
- Supply research for designing an Organizational Alignment architecture, making key strategic choices, selecting suitable software, or building a corporate memory.
- Garner support for new initiatives designed to improve the knowledge assets.*

The challenge that modern organizations face is to turn the scattered knowledge of their intelligent agents who are working in virtual teams into a well-structured knowledge repository.

—Alberto Cameiro, Associate Professor, University of Lisbon

You will note that knowledge inputs can be both tacit and explicit. A typical tacit knowledge input would be advice from an expert. A typical

---

* Ibid.

explicit knowledge input would be a design specification. Also the output from any activity can be either tacit or explicit. A typical tacit output from an activity would be lessons learned, and a typical explicit output would be a change request. During Activity 2, Analyzing the Map, the individual or team will ask questions like the following:

- What knowledge is missing?
- What knowledge is most valuable?
- What knowledge can be used in other processes?
- What knowledge is generated but not shared?

During Activity 3, Applying the Map, the following activities should be accomplished:

- Defining intellectual assets
- Developing a knowledge management system
- Improving knowledge flow
- Setting up knowledge management activities within the business
- Identifying holes in the restructuring plan

Outcomes from the Knowledge Mapping Process

There are some major outcomes involved in the knowledge mapping process. Typically, knowledge management systems are usually abstract, overly systematized, and weak in implementation. Using process knowledge mapping provides a concrete and tactical way of understanding knowledge and transforming it from hard into soft knowledge.

Key Processes and Network Implementation Challenges

We have found that the total delivery system must be put under the microscope if knowledge mapping is to succeed and significant changes are made in the value chain. What makes the mapping process so powerful is the blending of the various components into a synergistic whole. Identifying key influencers may be equally as effective in getting an organization properly aligned to its strategy as a major restructuring. Key influencers are employees who may not have formal authority or an official title, but they do have the ability to greatly influence their peers. Yet, time and again, when change is at hand, senior executives often keep these same employees "in the dark" because they are not part of the official hierarchy of the organization.

A Knowledge Management paradox* states, however, that the immense benefits do not directly translate into business value. An estimated total of 50% to 75% of all reengineering projects have failed, which is comparable to the TQM failures of the early 1990s. This paradox is summarized in a report by McKinsey and Company on business process knowledge mapping in 100 companies.† This is why knowledge mapping, in conjunction with Process Redesign, is used so extensively. It is over 80% effective at reducing cost and cycle time by 30% to 60%. By using knowledge mapping alignment principles, BPI evolves into a fundamental realignment of operating processes and organization structure, focused on the organization's core competencies to achieve dramatic improvements in organizational performance. In other words, it is a process by which any organization can realign the way it does business to maximize its core competencies, of which a basic mapping competency is critical to success.

Knowledge mapping helps to redesign the way work flows through an organization, often leading to system and infrastructure changes. It differs from traditional process mapping by focusing on the core competencies, or key factors for success.

> Select the right process and then select the right improvement methodology. Fail at either, and your project will fail.
>
> —HJH

---

* See Harrington and Voehl, *Knowledge Management Excellence* (Chico, CA: Paton Press, 2007).

† In all too many organizations, reengineering has not only been a great success but also a great failure. After months, even years, of careful redesign, these organizations achieve dramatic improvements in individual processes only to watch overall results decline. Managers proclaim a 20% cost reduction, and a 25% quality improvement yet in the same period business unit costs increase and profits decline. So we must ask the question: is this failure? Or does the problem lie with the process itself, which in many cases is never really implemented. In other words, they talk about it but rarely actually do it.

FIGURE 5.3

Hidden office. RVA = Real-Value-Added; BVA = Business-Value-Added; NVA = No-Value-Added.

SUMMARY

You should always design your processes to minimize hand-offs and cycle time. Eliminate the No-Value-Added (NVA) Activities and minimize Business-Value-Added (BVA) Activities. The No-Value-Added and Business-Value-Added Activities make up a huge hidden office that provides nonproductive services to your customer. This huge hidden office is costing 50% to 75% of all your efforts and cost. Do you know what it is costing you?

> Are you spending most of your money on Real-Value-Added (RVA) Activities? If you think you are, well think again.

—HJH

6

Phase III. Organizational Structure Design

Too many organizations reorganize for political not performance reasons.

—**HJH**

Phase III of the Organizational Alignment Cycle is upgrading the organization's structure (Figure 6.1).

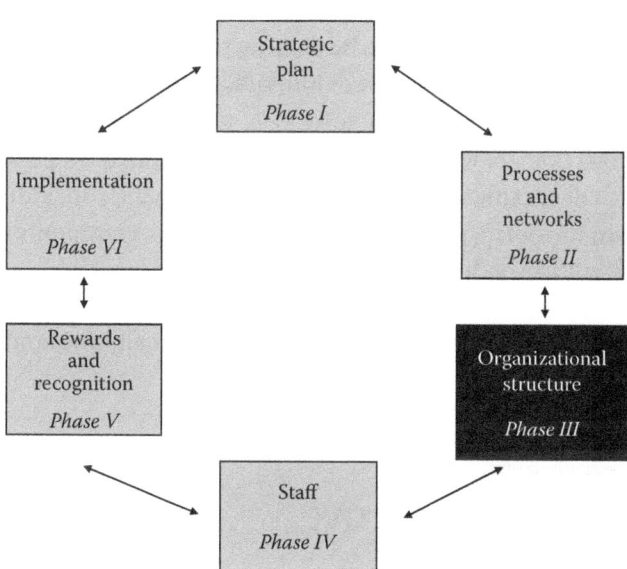

FIGURE 6.1
Phase III of the Organizational Alignment Cycle.

INTRODUCTION TO ORGANIZATIONAL STRUCTURE DESIGN PHASE

Organizational Structure Design is used to change the way an organization conducts its affairs in order to increase productivity, performance, and profit by providing the knowledge and methods to handle the kind of recurring organizational change that all businesses face, especially those which do not involve transforming the entire organization. Change generally necessitates significant change at the business unit, divisional, functional, facility, or local levels. The problem lies in knowing what needs to change and how to change it.

> Definition: *Business Unit* is a part of a larger organization that is directed at a unique product, service, or region. It usually is accountable for its own profit and loss performance.

Taking the organization as a "designed system," there are four major elements of Organizational Alignment:

1. The work—the basic tasks to be done by the organization and its parts
2. The people—characteristics of individuals in the organization
3. Formal organization—structures, i.e., the hierarchy, processes, and methods that are formally created to get individuals to perform tasks
4. Informal organization—emerging arrangements, including variations to the norm, processes, and relationships, commonly described as the culture or "the way we do things round here."

The way these four elements are aligned (relate, combine, and interact) affect productivity, performance, and profit.

THE STRATEGIC PERSPECTIVE

Why do businesses consistently fail to execute their competitive strategies? Because leaders don't identify and invest in the full range of projects and programs required to align the organization with its strategy. Moreover, even when strategy makers do break their plans down into doable chunks, they seldom work with project leaders to prioritize strategic investments and ensure that needed resources are applied in priority order. They often neglect to revise the strategic portfolio to fit the demands of a dynamic environment or to stay connected to strategic projects through completion, as new products, services, skills, and capabilities are transferred into operations.

The goals of the aligned organizational structure are to achieve better coordination between functions, decentralize authority, increase employee involvement at lower levels of the organization, restructure the decision-making process, more clearly define responsibilities, and encourage more innovative forward thinking. In this chapter, we present various imperatives that enable you to first design* and then do the right strategic alignment projects—and do those projects right. And it is no accident that the six imperatives combine to create the acronym INVEST:

Ideation: Clarify and communicate Purpose, Identity, and Long Range Intention.

Nature: Develop alignment between Strategy, Structure, and Culture based on Ideation.

Vision: Create clear Goals and Metrics aligned to Strategy and guided by Ideation.

Engagement: Do the right projects based on the Strategy through Portfolio management.

Synthesis: Do Projects and Programs right, in alignment with Portfolio.

Transition: Move the Project and Program outputs into Operations, where benefit is realized.

The world is increasingly becoming an online village open 24 hours, seven days a week.

—Jonathan Groucutt, Professor, Oxford Brookes University

RESTRUCTURING PRINCIPLES

An organization can be defined around six core concepts:

1. Organizational environment
2. Strategy

---

* For further analysis of these comments see Theodore Cohn and Roy Lindberg, *Survival and Growth: Management Strategies for the Small Firm* (New York: AMACOM, 1974); Peter Lorange and Richard Vancil, "How to Design a Strategic Planning System," *Harvard Business Review* (September–October, 1976).

3. Technology
4. Social structure
5. Culture
6. Physical structure

The environment, which lies outside the boundaries of the organization, provides inputs (resources) to the organization and influences its outcomes. Managing the environment-organization relationships is the main focus of top management. Figure 6.2 shows the interaction between the organization's components and the environment in which it operates.

Organizational structure refers to the relationships among the parts of an organized whole. There are two types of structures: physical and social. Physical structure refers to relationships between the physical elements of an organization, such as its buildings and the geographical areas in which it conducts its business. Social structure refers to relationships among social elements, including people, positions, and the organizational units to which they belong (departments, sections, units, etc.). An organization is a social structure (bureaucracy) consisting of a hierarchy of authority, a division of labor, and formal rules and procedures (Max Weber).

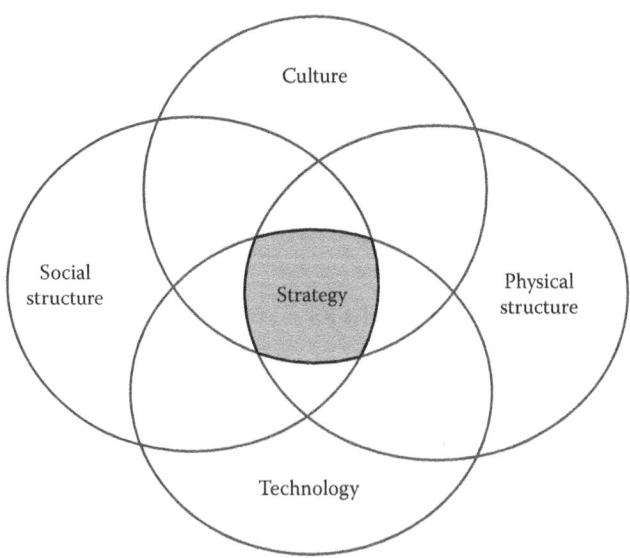

FIGURE 6.2
Organizational environment.

Definition: *Hierarchy* reflects the distribution of authority among organizational positions. Hierarchy defines formal reporting relationships that map out upward communication channels through which management expects information to flow.

Definition: *Authority* is the granting of the position holder certain rights, including the right to give direction to others and the right to reward and punish. These rights are called positional powers because they belong to the position rather than to the position holder.

Definition: *Division of Labor* defines the distribution of responsibility (while hierarchy defines the distribution of authority). Division of labor concerns the way jobs are grouped into organizational units.

In studying organizational structures a number of central issues, such as span of control, conflict, decision making, approval layers, power, and politics, have been defined as important. Figure 6.3 indicates how some of the dimensions considered are measured.

Organizational structures vary according to the requirements of organizations. Realignment principles are concepts that are used to guide the organization alignment team when they are considering different organizational structures. There are many different types of organizational structures. These include Decentralized Structure, Functional Structure, Geographic Structure, Product-Based Structure, Matrix Structure, Vertical Structure, Customer-Based Structure, Front-Back Hybrid Structure, Bureaucratic Structure, Network Structure, and more. Each of these structures has advantages and disadvantages, which must be carefully analyzed before an alignment design is chosen.

Dimension	Typical Operations Measured
Size	Number of employees of the organization
Administrative Component	Percentage of total number of employees with administrative responsibilities
Span of Control	Total number of subordinates over whom a manager has authority
Specialization	Number of specialties performed with the organization
Standardization	Existence of procedures for regularly recurring activities
Formalization	Extent to which rules, procedures, and communications are written down
Centralization	Concentration of authority to make decisions
Complexity	Vertical differentiation—number of hierarchical levels; horizontal differentiation—number of units within the organization (departments, units, etc.)

FIGURE 6.3
Dimensions of social structure.

Think the unprecedented number of site transactions is reserved for large companies? Small and medium sized companies are getting into the act, too.

—Carol Bergeron, founder, Bergeron Associates

Restructuring principles can be applied to any or all of the standard organizational structures. The restructuring principles that should be considered are the following:

1. Bureaucracy Reduction—The mission statement for each level within the organization should define the manager's authority level, thereby reducing bureaucracy. By reducing the number of management levels, bureaucracy is also reduced.
2. Reducing Impact of Bottlenecks—The design will consider how the negative impact of these bottlenecks can be reduced. One way is to eliminate the organizations or activities that are defined as bottlenecks and integrate their activity into the organizations they service.
3. One-Stop Shopping—The ideal customer service system would be to have the customer go to only one counter and interface with only one person. This one person will handle all the transactions for the individual. Even if the customers' needs cross sector boundaries, this one person would also collect the money for the service. Organizational designs that do not meet this process description are compromised designs.
4. Empowerment—Empowered people are happy employees and have highly satisfied customers. Typically management does not empower lower management or their employees to handle the situations they come in contact with because they do not trust the lower level managers or employees to make the proper decision or the individual will take advantage of the organization. Unfortunately, in most organizations, there are employees who cannot be trusted. We estimate that this is about 0.1% of the total population. As a result, organizations have established procedures and policies that are designed to prevent this 0.1% of the population from taking advantage of the organization and have punished the other 99.9% of the employees that are honest and trustworthy because management has not been able to identify those employees that would take advantage of the organization.

5. Decentralization—Often an organization may need to establish a number of satellite operations to bring services closer to customers. These satellite operations need to be manned and trained to service the needs of their geographically located customers. It would be desirable to complete all transactions for their customers at the satellite location without the customer going to the main office.

Informal communication, one of the most powerful tools in building an effective team, is harder to achieve in geographically dispersed teams.

—Elaine Crowley, founder, The Crowley Group

6. Outsourcing Non-Core Processes—No organization can excel in all areas. Those who try usually sacrifice the performance of their core business. Best practices today indicate that organizations should outsource their non-core processes to people who specialize in these processes and can perform them at lower cost, reduced cycle time, and higher quality.
7. Flattening the Organization—Improved communication and rapid decision-making processes are key elements to organizational excellence. Flattening the organization drives both of these key elements by reducing the number of management levels, costly bureaucracy, and cycle time while improving quality and employee satisfaction.
8. Span of Control Guidelines—Having too many employees reporting to a manager burns out a good manager and results in inadequate employee guidance and development. This often leads to missing schedules, increased cost, low morale, and decreased customer satisfaction. On the other hand, having too few employees reporting to a manager results in loss of management productivity and limits the growth of the employee because of the manager's involvement in the technical decisions.
9. Reduced Cycle Time—As good, effective managers, we are interested in minimizing process time. But from our customers' standpoint, they do not see processing time; rather, they measure cycle time, how long from the time they request a service was the service provided. Therefore it's extremely important for us to focus on minimizing cycle time by combining activities and minimizing transportation delays in order to achieve high levels of customer satisfaction.

10. Privatization—Privatization is very different than outsourcing. Privatization is selling part of the organization's activity to a private organization where outsourcing is defined as contracting part of the organization's activity to a private organization. In the case of outsourcing, the organization is still responsible for the results. When an activity or process is privatized, the organization is no longer responsible for the success or failure of the activity. As a result, privatization may be preferable because it immediately produces cash flow and removes the responsibility for performance from the organization.

11. Not Influenced by Politics—It is extremely important that the final organization design reflects best practices rather than the political situation facing the leadership team. Truly the culture of the organization has to be considered, but the design cannot be influenced by individuals who are worried about maintaining their turf.

12. Minimizing Movement—Every time an item moves from one desk to another, there is a delay and additional cost involved. When an item has to move physically or electronically to a different location within the building or outside the building, this delay is magnified. Minimizing document movement is a crucial item in reducing cost and cycle time.

13. Simplification—Too often we try to solve our problems by the use of information technology, when there are other less costly and less disruptive approaches that can be applied. Simplifying the way we process information and documents often results in bigger improvements and costs less than computerization.

14. Telecommuting—The world has accepted that in many cases individuals can work from their home more effectively and efficiently than in the work environment. Some of the work that is being done in the service groups is the classification of candidates to be handled remotely saving travel time and space. Organizations like Ernst & Young, which have gone to telecommuting, have reduced their occupancy space by as much as 70% while improving the quality of their output and the morale of their employees. We realized that this may be a radical thought pattern for many organizations but it is one that should be considered.

15. e-Business—Information technology can be very effective at eliminating the processing of routine tasks. It is great at saving money and reducing cycle time. Its impact on the organization needs to be considered in the final organizational design.

DESIGN CRITERIA

The organizational structures' design criteria are all based on the requirements set forth in the Strategic Plan. Typical criteria are given in Figure 6.4.

REALIGNMENT PRINCIPLES AND DESIGN CRITERIA

As organizational structures change, they can have both positive and negative impacts on the key performance indicators. Figure 6.5 explains these interrelationships.

List of Activities to Help Define the Design

The following is a list of things that should be done to collect the background information needed to do the organizational restructuring.

A. The new structure should be in line with the Strategic Plan.

B. There should be improvement in customer satisfaction.

C. It should improve internal customer satisfaction.

D. The new structure should improve decentralization.

E. The new structure should increase profit.

F. The new structure should enable the organization to delegate decision making to lower levels.

G. Each part of the organization should be designed based upon the organization's need and not upon an individual's need.

H. There should be different titles and job descriptions for technical leaders (lead experts) and individuals that manage people.

I. All organizational units' span of control should be within the span of control guidelines.

J. Non-core activities should be outsourced to reduce cost and improve quality, not to reduce head count.

K. The new structure should improve communication.

L. It should reduce cycle time.

FIGURE 6.4
Summary of design criteria.

Organization Type	Output Quality	Flexibility	Knowledge Mgmt.	Comm.	Cycle Time	Employee Satisfaction	Customer Satisfaction	Cost	Risk
1. Functional (Vertical)	+	–	–	–	–	+	–	+	–
2. Product / Customer	+	–	–	–	+	+	+	–	+
3. Matrix	+	+	+	+	+	+	O	–	+
4. Hybrid	+	+	O	+	O	–	O	+	O

Span of Control Guidelines with Improvement	Span of Control Guidelines without Improvement
CEO = 4 to 10	CEO = 2 to 6
Vice President = 4 to 10	Vice President = 2 to 8
Middle Manager = 4 to 10	Middle Manager = 2 to 8
Dept. = 6 to 12	Dept. = 5 to 8
Sectors = 8 to 18	Sector = 5 to 8
Units Prof = 8 to 18	Unit Prof = 5 to 10
Units Labor = 10 to 30	Unit Labor = 5 to 15
(+) Positive impact; (–) Negative impact; (o) Neutral.	

FIGURE 6.5
Realignment principles and design criteria.

1. Review the organization's 5-year plan to determine additional skills and work load requirement.
2. Meet with Personnel to define what job responsibilities relate to each level of management (CEO to first-line manager or foreperson) and work load.
3. Meet with each vice president and many other managers to define the present manpower inadequacies.
4. Review the KPI to determine capabilities meeting requirement with present personnel.
5. Review customer survey to define weaknesses.
6. Review job description to define technical requirement.
7. Review status of the 5-year plan with the Executive Team to ensure that it is on schedule.
8. Meet vice president/manager to get input related to Organizational Alignment.
9. Conduct a 2-day off-site meeting to brainstorm Organizational Alignment ideas.
10. Conduct an interface survey between all departments to identify primary interface and level of performance.

11. Flowchart the major process to understand relationship between departments.
12. Prepare a knowledge map to define interaction flow between departments.
13. Conduct a change management survey to define risk areas in making changes.
14. Conduct a coordination survey to define percentage of time each department spends interfacing with each of the other departments.
15. Consider the impact of each department on the general public.
16. Consider the impact of each department on the external customer.
17. Meet with customers to gain first-hand knowledge of the organization's interface and conduct a customer survey of a small sample.
18. Meet with a sample of alliance partners to provide input on alliance partnership activities.
19. Conduct an audit to define wait time for customers.
20. Prepare an investigation of the outsourcing strategy and plan.
21. Meet with the remote location to gain understanding of why their customers need to go to the central location.
22. Define what the organization's core capabilities and competencies are and focus on strengthening those key activities.
23. Consider the capital investment assigned to each area to ensure high investment areas were properly addressed.
24. Identify key business revenue generated points that have potential for increased revenue and include that in the design.
25. Identify weakness in the actual organization where departments need to be more pro-active and visionary.
26. Consider the work load in the Executive Team and try to balance it.
27. Minimize the number of levels between employees and the president or CEO.
28. Consider the impact of removing people from management and the impact it would have on the morale of the organization.
29. Consider the impact of transforming the personnel department into a HR function.
30. Consider the personality traits of the Executive Team.
31. Include the president's organizational structure desires.
32. Identify organizational bottlenecks and try to eliminate them.
33. Consider how case management approaches would be used for the alliance partners.
34. Consider how a one-stop shopping concept could be used.

35. Consider how the project could be managed better.
36. Consider the present level of the organizational units and the impact if it was downgraded.
37. Consider the long-range impact of outsourcing on the organizational structure. (Example: What if payroll was all outsourced?)
38. Consider the long-range impact of disruption to the job assignment at the employee level.
39. Determine how to implement without disturbing services or customers.

MEASUREMENT SYSTEM

The organization should have developed an extensive list of Key Performance Indicators (KPIs) that are updated and reported on a regular basis. Although the present system may be working well, there is always room for improvement. Four areas, in particular, need to be addressed.

1. Are all of the stakeholders' interests, requirements, and expectations covered?

    Stakeholders include government agencies, the community, strategic partners, outside investors, suppliers, consultants and contractors, management, employees and their families, and special interest groups. Each of these stakeholders has a vested common interest in the development of the organization, but they all have different perspectives and special needs that the organization must address. In a truly balanced scorecard, all the positive and negative impacts that an organization has on all stakeholders should be addressed.

2. It is important to understand the changing needs of each stakeholder and to set KPI targets and measurements to reflect these needs. This can only be done through regular consultations with individual stakeholder groups, which should lead to the development of clear requirements and measurements. Currently most organizations are focused on measuring external customer satisfaction, but it is not doing enough to measure and report on internal customer satisfaction. An obvious weakness in most present systems is defining acceptable performance criteria for the support organizational units, as they relate to the operational areas.

3. Are all of the major processes performed by each level covered?

At each organizational unit level (sector, department, section, unit, supervisor, and foreperson) there is a group of processes that the natural work teams are involved in. Processes that consume more than 10% of an organizational unit's resources can be considered major processes for that unit. This means that no more than ten major processes per area can be defined and usually there are only three to six. For each of these processes, effectiveness and efficiency measurements should be defined. This type of process focus has not been applied sufficiently to the measurement system in most organizations. By using this approach, the organization will greatly improve the overall effectiveness of the measurement system.

4. Is there a direct correlation between the KPIs and the Strategic Plan?

It is good practice to relate each major improvement program to a resultant measurement. It is suggested that each organizational unit responsible for each strategic goal or objective document the KPIs that each activity will impact.

The Strategic Plan has placed increased emphasis on the impact that the organization processes will have on all of the stakeholders. There needs to be an increased focus on measuring the quality of the work that is performed. The survey we conducted indicated also that there is a need in most organizations to have a better system to measure the quality of the services provided to internal customers. This does not suggest that there should be any relaxation in the present measurement system, as it is needed to continue to support the present performance.

Key Performance Indicators

We suggest that a Balance Scorecard approach be used in the future to measure the impact the reorganization structure will have on the organization. The following six measurements are typical ones to be monitored.

- **Customer Satisfaction**—This measurement is one of the KPIs, and an extensive database should be collected to define the as-is situation.

The idea of loyalty to one, and only one brand is a total myth.

—Jonathan Groucutt, Professor, Oxford Brookes University

- **Employee Satisfaction**—This is a calculation that is done related to the HR department.
- **Communication Index**—We often find that there is no KPI that measures this particular consideration adequately. If that is the case, a communication effectiveness survey should be conducted and these data will be used as a starting point. Repeating this survey after the organizational structure design has been implemented and emotions have stabilized will provide the organization with a Gap Analysis.
- **Operating Cost**—The reduction in the total budget is an effective way to measure this.
- **More Effective Implementation of the Strategic Plan**—This item can be measured by measuring the percentage of the Strategic Plan activities started in the past and completed on schedule compared to the percentage of Strategic Plan activities started after the new organizational structure design has been implemented and completed on schedule.
- **Improved Output Quality**—These measurements will vary based upon the products and/or services that are provided. Certainly, customer satisfaction is one indication. The other indication that we propose is measuring the number of customer complaints per total population.

It is important to realize that reorganizing usually has a negative impact on overall performance for a period of time. As the organization stabilizes, the improvements will start to take hold. This is caused by people going through a learning curve related to their assignment and how they need to interface with their management team. Those measuring the impact of the new organizational structure need to take into account this condition.

RESTRUCTURING APPROACH

Our approach shows you how to align and make strategy happen in your organization, based upon both recent and vintage studies going back as much as 50 years or so.* Regardless of whether an organization is trying to radically improve its customer relationship management efforts or simply create a better way of measuring one component of performance, the answer to improvement seems to lie in how often an organization induces change and how sweeping that change is across its component organizational levels. Organizations that are effective at getting aligned behind its goals do two things:

1. They tend to focus on moderate, intermittent changes that alternate between a period of disruption and a period of stability.
2. They look long and hard at their existing organizational assets to understand how they can be repurposed, redeployed, or combined differently to meet the new goals.

When an organization decides to undertake a disruptive effort, it needs to think through whether or not its employees are already suffering from initiative overload. Also, executives must honestly ask themselves if the focus is on key changes that will really move the organization closer to its goal. Because there is usually a state of disorder between when the change is initiated and when it is completed, employees are in a state of flux and are doing things that require more effort and energy than usual.

Launching another big effort before the previous one gets adopted creates exponential chaos and disorder. In fact, it is not the outcome of change that is not desired by employees, it is the disruption and uncertainty that require energy to push through—too many of these initiatives without a period of stability in between will have your best employees heading for the door.

---

* The first comprehensive and documented study was Stanley S. Thune and Robert J. House, "Where Long-Range Planning Pays Off," *Business Horizons* (August 1970). In their study "Does Planning Pay?" *Long Range Planning* (December 1970), H. Igor Ansoff et al. determined that those organizations that merged with others on the basis of long-range Strategic Plans did better than those that did not plan. However, another study by Robert M. Fulmer and Leslie W. Rue, *The Practice and Profitability of Long-Range Planning* (Oxford, OH: Planning Executives Institute, 1973), raised some doubts about the benefits of Strategic Planning in service industries. One of the more interesting studies was a doctoral dissertation that examined the experience of 90 organizations in the United States in different industries, which ultimately concluded, on the basis of financial measures, that those organizations that had long-range planning systems did better than those that did not. The results were first summarized in Delmar Karger, "Long Range Planning and Organizational Performance," *Long Range Planning* (December, 1975) and were systematically reviewed and updated some 30 years later.

The Roadmap for Organizational Restructuring

The Harrington Institute Roadmap for Organizational Restructuring and Design (HI-ROAD) clarifies why and how organizations need to be in a state of readiness to design or redesign and emphasizes that people as well as business processes must be part of design considerations. If you're thinking about redesigning your organization—remember, it's all about aligning with your strategy. Strategy—which includes vision, mission, and goals—should be the foundation for building your organization, and it should be front and center every time you make a decision about your organization. Need to decide what your structure and reporting lines should look like? Revising your reward system? Considering how to attract, retain, and develop the right people? Trying to foster better collaboration and networking across groups? If you're tackling any of these challenging issues, you should refer to your strategy as a guidepost for making organizational design decisions.

The process of design must be aligned with objectives. If the objective of design is a system capable of self-modification, of adapting to change, and of making the most use of the creative capacities of the individual, then a constructively participative organization is needed. A necessary condition for this to occur is that people are given the opportunity to participate ultimately in the design of the jobs they are to perform.

> Those whom we would identify as "creative" are so because they provide themselves ... with a huge array of stimulus material for sparking new ideas, drawn from their total life experience.
>
> **—Jeffrey A. Govendo, founder, The Innovative Edge**

THE ALIGNMENT PROCESS

We use a five-activity approach anchored by an effective front-end assessment, followed by analysis and redesign, supported by documentation, and ending in a successful implementation, as shown in Figure 6.6.

> We might as well require a man to wear the coat which fitted as a boy.
>
> **—Thomas Jefferson**

When Thomas Jefferson made this statement, he was discussing the need to reorganize the way governments were structured. His thoughts are still

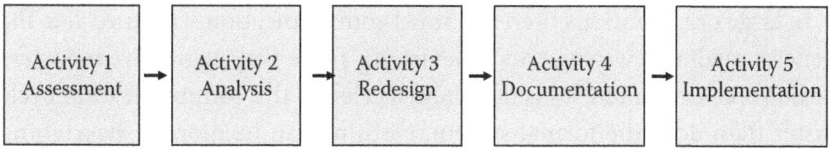

FIGURE 6.6
The alignment process.

true for both the public and private sectors. As our organizations grow, we can let out the seam of our organization just so much and then we need to realign the organizational structure. In these cases, a holistic approach is required. This section discusses the things that should be considered and how to go about making a total assessment that results in a redesigned organization structure.

Activity 1. Assessment

During the assessment activity, the following activities take place:

- Assess existing organizational structure
- Identify improvement opportunities
- Evaluate resistance to change
- Evaluate the measurement system
- Assess alignment between strategy and culture
- Benchmark Studies
- Communication Study
- Interfaces and Coordination Studies
- Customer Survey
- Key Performance Indicators Analysis
- Organizational Restructuring Workshop

Communication Maps

The basic foundation of an organization is its *communication structure*. In organizations that are made up of two or more people, there are at least two communication structures—formal and informal.

Usually the informal communication structure is faster than the formal one. The old rumor will get the word around faster than the electrons that can travel through the computer lines. The problem with the informal system is that it often provides misleading information about individuals and the organizations.

In large organizations there is a third communication structure. It is the general media (newspaper and television). These very aggressive reporters of hot-breaking interest events often accelerate the communication cycle faster than does the formal system. Nothing can be more disheartening than to hear something on the radio or to read about in the newspaper when it affects your job and you have not been told about it before the rest of the world became informed.

One way to analyze and to communicate an organization's communication structure is to develop maps of the important communication system for the new organization's structure. Figure 6.7 is an example of a day-to-day communication map.

Interface Study

This study is designed to define the interdependence between operating units and assess the level of cooperation among the various units within the organization. The managers are asked to express their opinion about

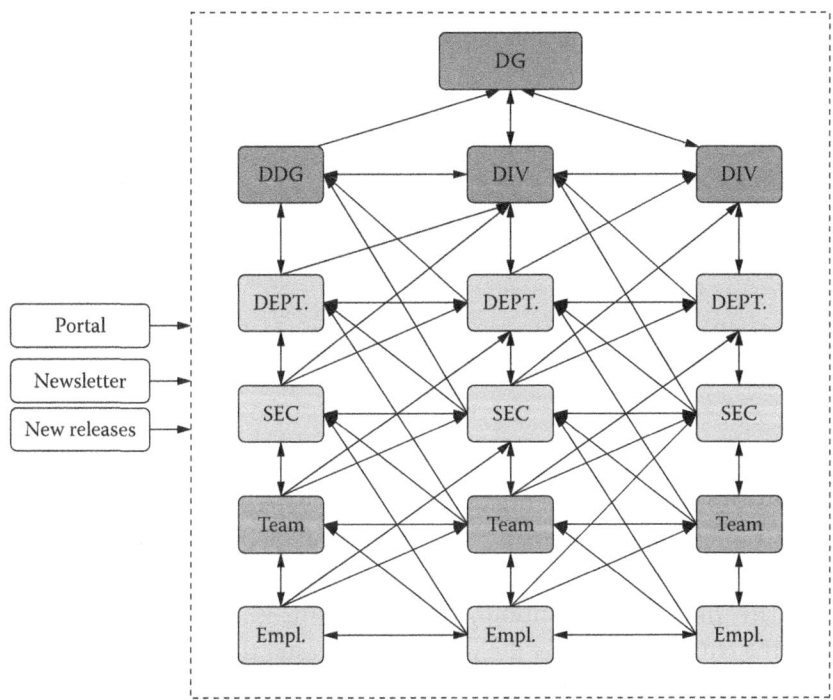

FIGURE 6.7
Day-to-day communication map.

the level of satisfaction they have with the service that they get from other units that they work with. A typical summary report is shown in Figure 6.8.

As you can see, this is a very effective way to define bottlenecks and improvement opportunities.

Organizational Restructuring Workshop

Restructuring is one of the most painful activities that an organization needs to undertake. The more people that get involved, the less painful it will be and the better it will be accepted. The other side of that is lower level people often get upset if they know the organization is going to be restructured and do not know how it will affect them.

One way that we have found effective in getting the project started is to hold an off-site Organizational Restructuring Workshop with the senior staff and executives. At their workshop, the "Organizational Structure Design Concepts" are presented. Then the attendees are divided up into small groups that help define the future state structure by working on exercises like:

1. What would the organization look like if it was flattened as much as possible to reduce cost and improve communications?
2. What organizational sections, departments, units, or sectors should be combined to reduce redundancies?
3. How would the organization need to change to eliminate or reduce identified bottlenecks?
4. What decisions are being made by the Executive Team that could be made at a lower level?

Unit	Rating	Number of Responses			
		Very Disappointed	Disappointed	Satisfied	Very Satisfied
A	2.06	6	18	8	0
B	2.25	2	8	6	0
C	2.29	4	16	10	2
D	2.29	8	10	14	2
E	2.31	6	8	10	2

FIGURE 6.8
Low end of the interface summary report.

Activity 2. Analysis

During the analysis activity, the following activities take place:

- Conduct change management studies focusing on culture, capability, communications, consistency, cooperation, career development, competencies, and competitiveness.
- Develop a change management plan.
- Develop an organizational restructuring guideline.

There are many reasons why individuals and organizations resist change. Some of them are:

- Lack of vision on the part of the individual and understanding where the organization's vision is leading
- Poor implementation history when changes have been tried in the past
- Lack of middle management support of the change as middle management tendencies are to resist change
- Lack of understanding or belief that change is necessary
- Low risk taking in the organization
- No consequence management system in place
- Lack of clear communication
- Failure to anticipate resistance
- Poor management of resistance
- Lack of time to understand the need for reorganization
- Poor follow-through when activities have been defined
- Lack of synergy throughout the organization

The Strategic Planning process leads you through a series of questions to determine how to best allocate and focus resources to carry out the strategy of the organization and achieve the best outcomes. To start this process, you should develop an understanding of the current state of the organization and its environment. Some questions that your leaders should explore are the following:

- How does the competitive environment shape the way we conduct business? Is a specific structure forced on us by competitors?
- How well do we meet our customer's demands? How does structure affect our ability to meet customer demands?

- What are the interrelationships among the different functions and units and how do they impact each other? How can structure support the coordination among them?
- What are our core competencies? How well do they support our strategy? How can structure facilitate the strategy?
- What impact does our history and culture have on how we have structured our organization? What barrier could it impose on a new structure?

The answers to these questions will have a dramatic impact on what you can and should do with your structure. For example, if reacting to a volatile market where the customer needs are constantly changing is the number one strategic issue, then a decentralized product design unit attached to a horizontally based organization makes more sense than a centralized design center at corporate headquarters. One way to focus the effort of answering these questions is to approach the processes by looking at data input, influencing principles, and output options. Figure 6.9 is an illustration of how one organization used that approach. This process is normally conducted over a period of weeks during which information is collected on the data inputs. That is then presented and filtered through the principles with the resultant outputs used to help set the direction for the next activity which is to address the operational perspective.

Data Inputs	Principles	Output Structure Options
Market Trends		
Customer Needs		
Competitors	Level Management	Vertical
Business Performance	Customer Oriented	Matrix
Suppliers	Growth vs. Cost	Network
Industry Trends	Employee Involvement	Case Management
History	Process Orientation	Horizontal
Strategic Plan	Market Oriented	Process Management
Organizational Values		
Employee Skills		

FIGURE 6.9
Organizational structure visioning matrix.

Activity 3. Redesign

The structure of an organization refers to the formal way in which people and work are grouped into defined units. Any organization with more than two dozen people or so will need to begin to group people together in order to manage the work effectively. Grouping activities and positions into organization units establishes a common focus by creating standard processes, access to information, and a common chain of authority. It allows for efficient use of organizational resources and provides employees with an identifiable "home" within the larger organization.

The structure sets out the basic power relationships in the organization—how limited resources such as people and funds are allocated and coordinated. The structure defines which organizational components and roles are most central for execution of the strategy and how the business's profit centers are configured. No one structure is best for every organization. The best structure is the one that helps the organization achieve its strategy. There are multiple ways to structure the organization to achieve its goals. As with every design choice, each involves trade-offs and compromises. The objective in choosing a structure is to maximize as many of the strategic design criteria as possible, while minimizing negative impacts.

The predominant need regarding organizational structure is for a cross-functional focus—reducing the isolation of functions between divisions. The final organizational structure may require features of a horizontal organization in which structure is built around processes and teams.* This will include cross-functional teams focused on the four basic processes and a flatter organization to drive decision-making authority to a lower level.

Not all structured visions need to be so complex. One organization did theirs just by developing some bullet points:

- Manage by process rather than function.
- Achieve a common goal of satisfying the customer.

---

* In the oft-quoted words and wisdom of planning pioneer Aaron Wildaysky: "Planning may be seen as the ability to control the future consequences of present actions. The more consequences one controls, the more one has succeeded in planning. Planning is a form of causality. Its purpose is to make the future different from what it would have been without this intervention. Planning therefore necessitates a causal theory aligning the planned actions with the desired future results. Planning also requires the ability to act on this theory; it requires power. To change the future, one must be able to get people to act differently than they otherwise would. The requirements of successful planning from causal theory to political power grow more onerous as its scope increases and the demands for simultaneous aligned action multiply at a geometric rate." See Aaron Wildaysky, "Does Planning Work?" *Public Interest* (Summer 1971): 101.

- Have departments interact with each other before making policy or procedure changes.
- Lessen finger pointing between departments and divisions.
- Give employees a chance to have an understanding of their own functions (job responsibilities) in respect to the entire system.
- Provide consistency in work between departments (i.e., every individual performing the same task in a similar manner).
- Help employees learn from each other.

The Operations and Tactical Perspective

Innovators look for what new things customers value, rather than focusing on differences among customers. Often organizations rely too much on market segmentation and forget that segmentation techniques work well only in stable settings. Segmentation is difficult to execute in a turbulent environment in which the value proposition constantly changes. In subtle ways, operational aspects of Organizational Alignment are fundamentally changing the customer value proposition. In recent years, value innovation across all service dimensions—speed, convenience, personalization, and price—has accelerated as a result of technological innovations such as the Web and e-commerce. These innovations have substantially changed the underlying value proposition, which in turn has changed the operational capabilities and competencies needed by organizations.

Purpose of the Perspective

The purpose of this perspective is to review the design impact of the strategic organizational level and how an organization can cluster its work to support the strategic intent and direction of the business. One approach is to use three grouping options. These are activity, output, and segment (see Figure 6.10).

Each grouping has relative advantages and disadvantages in terms of competitive response, market response, and internal functioning and strategy implementation. Dividing the organization by activity is similar to the traditional vertical organization in which activity is defined as a function or knowledge group. Such an organization would have predominantly functional components at the highest level such as finance, operations, sales and marketing, and so forth.

Divisions based on activity usually promote high functional expertise and utilize staff efficiently. This is particularly effective where functional

Grouping Option	Structural Implications	Example
Activity Function Knowledge/Skill	All personnel who contribute to or accomplish similar activities or who perform similar functions are grouped together.	Auto manufacturers have historically used activity as the primary method of grouping; i.e., marketing, manufacturing, and service were all separate divisions.
Output Product Service Project	All specialists needed to produce a given product, service, or project work together.	Auto manufacturers have moved to a more horizontal structure, which includes cross-functional teams by output. For example, there are different divisions for minivans, luxury cars, and compact cars.
Segment Market/Industrial Users/Clients Geography	All specialists needed to serve an industrial/market segment, or to meet user/client needs, or to serve distinct territories work together.	Banking typically is divided by region, with some divisions serving the east, central, and western portions of the country.

FIGURE 6.10
Analysis of the three grouping options.

expertise and knowledge transfer are key to a strategy. However, because the work process tends to run across divisions, interdivisional tensions are likely to be observed.

Approach for Designing an Organizational Structure

Once the decision to change your organization's structure has been made, the next question usually is to decide, "What is the best organizational fit for my strategy and competitive environment, which makes best use of our distinct core competencies?" The answer comes not from a single diagnostic tool but rather from a technique of "informed dialogue" which is a combination of analysis and dialogue conducted in an interactive way. What is the best way to decide on the "right" structure and fit for my organization? The first outcome of this process is the ability to look at the organization from a three-layered perspective:

- Strategic perspective
  The strategic perspective looks at the organization from the top down and determines the overall shape of the organization. It's a process of moving the big boxes around to determine the right fit.

- Operational perspective
  The operational perspective deals with the strategic organizations. In this case you look at the organization from two directions. You review the strategic fit with a look from the top down. You ensure the appropriate mix of operational, managerial, and support processes through a bottom-up review.
- Tactical perspective
  The tactical perspective is completed with a bottom-up approach and determines the work team and job designs.

All of these outcomes (strategic, operational, and tactical) comprise what is called organizational structure. It is the combination of strategic, operational, and tactical decisions that will be the basis for determining the "right" organizational structure.

Dividing the organization along output lines allows each product group to focus on the efficient production of a specific product/service. Such an organization would have predominantly product/service components at the highest level, such as consumer electronics, industrial products, warranty operators, and components. Divisions based on product/service usually promote increased product innovation and productivity advantages. They tend to provide a rapid response to existing markets. This is very effective in a highly competitive market where production efficiencies are key.

On the other hand, coordination of marketing activities across different product groups is generally less effective. Also, any leverage that may be achieved with supplier and distribution channels through coordinated purchasing and logistics is generally less than that of activity designs.

> Definition: *Organization Segments* are smaller subgroups comprising like or supporting types of activities. They may be divided by geographic, industrial/market segment or user client needs.

Dividing the organization along segment lines allows each group to focus on the responsive delivery of products and services. This method results in specific structures like the Americas group, European operations, high-net-worth clients, and so forth.

Divisions based on segments typically promote faster time to market and enhanced customer sensitivity and focus. They tend to have well-integrated customer support systems and rapid response to customer needs. They are highly conducive to a market where customized products or services are the norm. Most organizations use two or all of these groupings. (For example, often within traditional, activity-grouped organizations where the largest divisions are operations, sales, and service, subdivisions are usually based on output or segment.)

The key is in being able to identify the right combination leading to value innovation. Most of the answers to doing this will come from the strategic perspective process and the development of the structural vision. Remember that the operational perspective is the bridge between the strategy of the organization and the way in which the work is performed.

It's important to keep in perspective that the groupings are not mutually exclusive. In fact, the most effective organization uses all types of groupings. They may be grouped by activities at the senior management level, by market segment at the division level, by product at the plant management level, and by user at the work flow or process level.

There are 10 major organizational structure approaches and options. Each of them has advantages and disadvantages.

Option 1. Functional

Advantages

- Good career path
- Develops professional people
- High levels of capabilities in the profession
- Creates a body of knowledge
- Economies of scale
- Low level of duplication
- Easy to manage as the manager needs a narrower skill range
- Better standardization

Disadvantages

- Resists change
- Hard to manage diverse product line
- Priorities are often out of line with needs of the business
- Processes do not run smoothly
- Problem with cross-functional handoffs
- Longer problem-solving cycle

Option 2. Vertical

Advantages

- Very good in a stable environment
- Good transfer of priorities and knowledge
- Well-defined scope of activities
- Fewer handoffs

Disadvantages

- Narrow skill base
- Limited career growth opportunities
- Less professionalism
- Resistant to change
- Does not interface well with other functions
- Limited flexibility

Option 3. Bureaucratic

Advantages

- Excellent at the micro-level
- Total organizational focus
- Well-defined roles and responsibilities
- Stable quality and performance level
- Stable systems
- Effective strategy implementation

Disadvantages

- Inflexible procedures and policies
- Hard to change strategy
- Poor cooperation
- Internationally focused
- Potential for long cycle time

Option 4. Decentralized

Advantages

- Custom focused
- Good unit-level controls

- Organizations focus on success
- Organizations are self-sufficient
- Quick reaction to changing markets

Disadvantages

- Low level of organization focus
- High level of internal tension for resources
- Duplication of resources
- Hard for a organizational standard to be held to
- Poor knowledge transfer

Option 5. Product

Advantages

- Excellent team spirit
- High levels of technical skills
- Shorter development cycles
- Quick reaction to the competition
- Good accountability for performance and profits

Disadvantages

- Lack of standardization
- Reduced focus on total organizational goals
- Increased competition between product lines
- Reduced economies of scale
- Many different interfaces with the customer—no single contact point

Option 6. Customer

Advantages

- Strong supplier-customer relationship
- Customized products for the customers
- Fast reaction to changing needs
- Longer relationships

Disadvantages

- Many market segments
- Reduced economies of scale
- Duplication of activities
- Costly development cycles

Option 7. Geography

Advantages

- Direct focus on a specific region
- Better coordination within the organization
- Good performance accountability
- Close contact with customers
- Better customer understanding
- Fact reaction times

Disadvantages

- Poor resources sharing
- Difficult to mobilize
- Duplication of resources, both equipment and personnel
- Less specialization
- Poor handling of cross-region customers

Option 8. Case Management Network

Advantages

- Very responsive to customers' needs
- Total accountability for activities
- Very decentralized
- Requires flexible employees

Disadvantages

- Hard to stay strategically focused
- Requires more resources
- Not organization-performance focused
- Product and services performances can have a lot of variation

- People must be highly skilled
- Poor definition of roles and responsibilities

Option 9. Process-Based Network

Advantages

- Excellent alignment with the work flow
- Lowers cycle time and processing cost
- Good control over the required resources
- No functional boundaries
- Good accountability

Disadvantages

- Poor knowledge sharing
- Duplication of resources
- Limited career paths
- Reduced economies of scale
- Competing goals between process times

Option 10. Front-Back Hybrid

Advantages

- Ability for commodities product to provide value-added system and solutions
- Focus on product and excellence of product
- Variety of distribution channels
- One-stop shopping
- Cross-selling of products

Disadvantages

- Poor information sharing
- Complex accounting system required
- Conflicts over where resources are assigned
- Conflicting measurement systems
- Difficulty in coordinating marketing activities

Span of Control and Organizational Structure

Today, many senior managers are struggling with this question: "What kind of aligned architecture—cross-functional processes, integrated applications, and IT infrastructure—is needed to support new ways of doing business? And what is the span of control issues involved?" Most non-technology executives are flying blind in terms of emerging technologies at a time when it's necessary—even critical—to adopt them. Generally, they rely on their IT people. But senior executives who rely on IT managers to relate technology to overall business strategy do so at their own peril. Executives can eliminate their strategic blind spots by taking responsibility for understanding the implications of up-and-coming technologies and anticipating when they'll affect business strategy, including the span of control issues involved.

> Much is written about managing IT projects but little is written about managing a collection of IT projects.
>
> **—Vin D'Amico, founder, Damicon, LLC**

What Is the Correct Span of Control?

The following is a review of span-of-control alignment issues.

1. Managing too few people—When you eliminate the managerial assignment, you gain 50% of the individual's technical productivity that was previously dedicated to people management and business management.
2. Managing too many people—When you divide up a department that is over its recommended span of control, you increase the productivity of the people reporting to the manager by 10% to 20%. On occasion we have seen this go up as high as 30%.
3. A technical department that has 19 people in it with one manager does not allow the manager time enough to give proper direction and follow-up to the people he or she manages. If the department divides into two departments of 9 people each and one of them is made the manager, the productivity increases.
4. Assume that the productivity of the 18 non-managers increases by 15%, that is an increase of 2.7 employee days per day, plus .5 days of the new manager's time, for a total of 3.2 days per day increase in

productivity. That amounts to an 18% increase in output with the same number of people.

5. One of the objectives of most organizational restructuring is to reduce the number of levels of management. As management levels decrease, the effectiveness of communication improves as long as the span of control guidelines is not exceeded. We are always shocked when we find a significant number of managers at an organization with fewer than four people reporting to them. These managers end up doing the technical part of their assignment but fail at the business and personnel part of their jobs in most cases.

Balancing Managerial Work Load

When you are maintaining an organizational structure, it mainly involves balancing the work load of the individual managers; it is not one of redefining if a part of the organization should be at a lower or higher level. Normally a new organization is created when an individual manager's work load is too much for him or her to handle. A section could be divided into two sections if the work load was too much for the manager or the span of control was exceeded by more than 25%. For example, if the section manager's span of control grew to 14 professionals, it should be divided into two sections, 7 employees in each section.

> If the boss works seven days a week, cancels vacations, misses family events, and does not take time to exercise, subordinates follow suit.
>
> **—Barbara Kaufman, President of ROI Consulting Group**

The following are proven guidelines for proper span of control ratios by organizational levels in a typically aligned organizational design:

- Director General, Managing Director, CEO, President, etc. (Top Manager)—4 to 10 managers of sections, departments, sectors, or offices reporting to him or her
- Deputy Director General, Vice President, Deputy Director, etc. (Executive Management)—4 to 10 managers of departments, offices or sections reporting directly to him or her
- Section Managers, Functional Managers, Project Managers, etc. (Middle Management) Level 2 Management—4 to 10 sections or offices reporting to the Department Manager

- Section Managers (Professional Line Management)—4 to 10 employees; the majority of them are required to have a university degree. They are doing work that requires a high degree of advanced skills and creativity
- Units, Managers, Department Managers, Team Managers, etc. (Non-professional Line Management)—10 to 30 employees; the majority of them are not required to have a university degree. They need general direction and control as they are working using standard procedures. These individuals may be doing extensive physical labor and they usually are not required to make business decisions.

Team coordinators are not part of the official organizational structure but are often used when a group of people work together to solve a problem, to develop a plan, or to implement a project. The team coordinator can be appointed by the members of the team or management. The team may last for a few hours, days, or months. The team coordinator does not have any responsibilities related to how the individual team members perform.

Restructuring Challenges

As much as I would like to tell you that everyone should evolve to a Network Organization, I cannot. My experience indicates that all of the different organizational structures and even the combination of them must be considered depending on the organization's environment.

—HJH

The implementation challenges involving our five-activity approach follow a systematic logic of their own, and they're the same for everybody—startups, visionary and established organizations:

1. Challenge traditional definitions of value.
2. Define value in terms of the whole customer experience.
3. Engineer the end-to-end value stream.
4. Integrate, integrate, and integrate some more. Create a new techno-enterprise foundation that is customer-centric.
5. Create a new generation of leaders who understand how to create the digital future by design, not by accident.

These challenges are especially pertinent for established organizations, because they need the most help in transforming themselves. It is critical for established organizations to understand that we, as a nation and a world, are at a crossroads in history, a time when processes are making a transition from the fringe market, dominated by innovators and early adopters, to the mainstream market, dominated by pragmatic customers seeking new forms of value. Established organizations that don't pay attention to this shift and align their processes and systems accordingly are going to face hard times.

Restructuring Effectiveness

Customers want organizations that they do business with to continuously improve the following:

- Speed. Service can never be too fast. In a real-time world, there is a premium on instant, accurate, and adaptive response. Visionary organizations embrace constant change and consistently deconstruct and reconstruct their products and processes to provide faster service.
- Convenience. Customers value the convenience of one-stop shopping, but they also want better integration between the order entry, fulfillment, and delivery—in other words, better integration along the supply chain.
- Personalization. Customers want organizations to treat them as individuals. Artificial constraints on choice are being replaced with the ability to provide the precise product customers desire.
- Price. Nothing can be too affordable. Organizations that offer unique services for a reasonable price are flourishing, benefiting from a flood of new buyers.

Impact

The number one outcome is value innovation. Faced with similar products, too many options, and lack of time, the customer's natural reaction is to simplify by looking for the cheapest, the most familiar, or the best-quality product. Obviously, organizations want to locate themselves in one of these niches. A product or service that is 98% as good, isn't familiar, or costs 50 cents more is lost in a limbo of invisibility. Organizations that follow middle-of-the-road strategies will underperform.

Whether we are dealing with a team of people or an individual, we still need to determine how to structure work at a tactical level. As with the previous two perspectives, there are some questions to ask, which, depending on the answers, will give us some direction in this process.

1. To what degree does the job have a clearly identifiable beginning and produce a meaningful product or service? What is the interval of time between completing a task and the completion of the work process?
2. What categories of work logically are grouped together for an individual or team? What is the relationship each task has with the task preceding and following it?
3. How should work be managed and coordinated? Who makes the decisions? What level of decision making is made? How should decisions be made?
4. How do information and knowledge flow? Is feedback complete, immediate, direct, and individualized on tasks and operational completeness?
5. How routine is the work? Is it governed by standard methods and procedures? Does it change from day to day, customer to customer?

It is the cumulative knowledge of the three structural perspectives that finally provides you with the data necessary to put together a design plan for the organizational structure.

One good way to determine if your design will be a major improvement is to compare the present process to the future state solution. Coordination between units means integrating or linking together different parts of the organization to accomplish a collective set of tasks. This study is usually done before and after the organization is restructured. Figure 6.11 is a picture of the organizations that spend 10% or more of their time coordinating with the indicated organizations in the current structure. Figure 6.12 is the same analysis based upon the restructured organization's coordination flow.

Activity 4. Documentation

During the documentation activity, the following are developed.

- A detailed assignment description for each of the units
- Mission statements for each of the functions
- Organization chart for the organization's units down to the natural work team level
- Reporting structure is developed and communication linkage
- Revised organization manual
- Management-level authority and responsibilities matrix

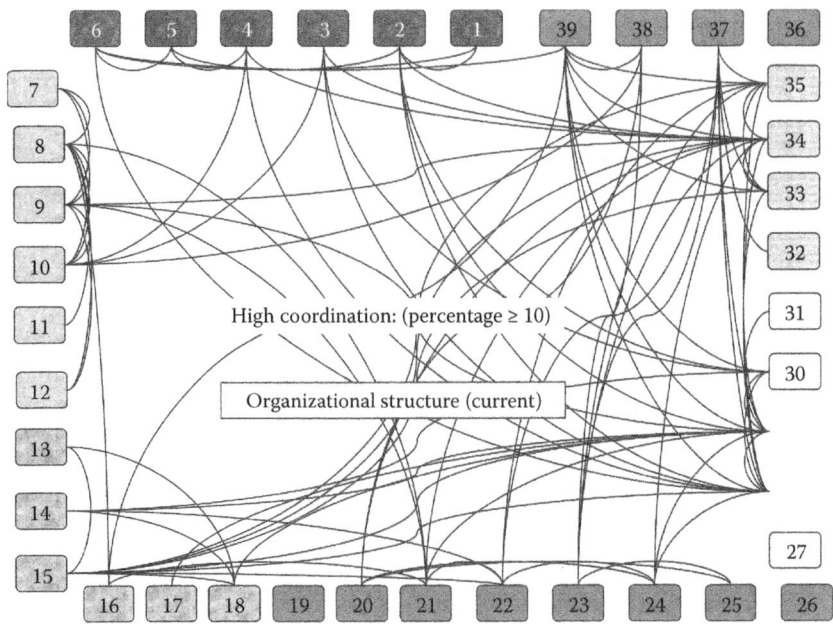

FIGURE 6.11
Organization structure (current).

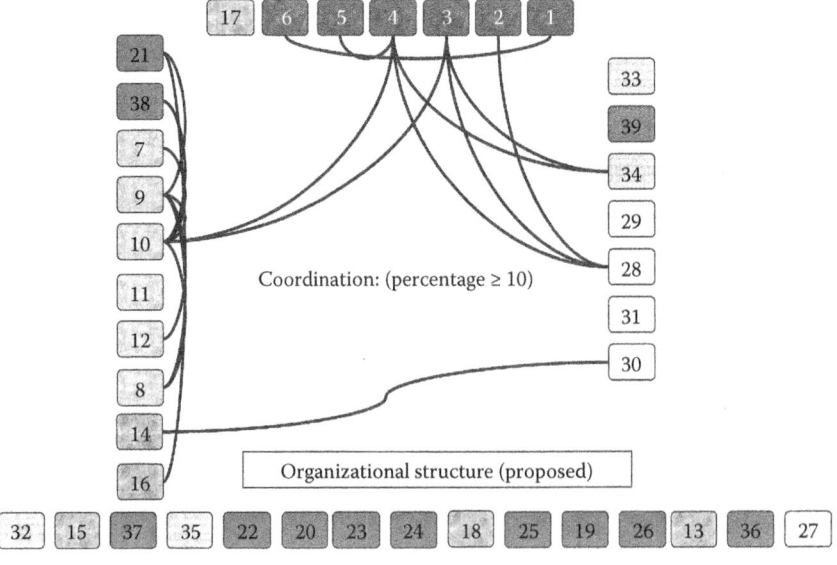

FIGURE 6.12
Organization structure (proposed).

- Flowcharts of all major processes
- A detailed implementation plan

We are not going to devote much space to this activity, not because it isn't important for it is very important and not because it is easy to do, for it is very hard to do well. We have decided to keep this discussion short because the documentation varies so much from organization to organization and it is really beyond the scope of this book.

There is one documentation approach that we feel we must bring to your attention. It is the management-level authority and responsibilities matrix. Figure 6.13 is the first page of one that we developed for one of our clients.

- ADG is the assistant director general or vice president's departments.
- Managers are functional managers.
- Section managers are project managers.
- Unit L3 are managers that manage degreed people or supervisors.
- Unit L4 are managers that manage non-degreed and degreed people who do jobs that don't require a great deal of creativity.
- Supervisors are people that have non-degreed people reporting to them.
- Forepersons serve as team leaders.

Activity 5. Implementation

The first consideration is to select the managers for each of the blocks in the new organizational chart. Often organizations make a big mistake by having employees assigned to a block on the organization chart that does not have a manager assigned to it. This is something that should never occur. If you don't have the right person to do the management job, don't assign an employee to the department or, at the very at least, assign an acting manager who will manage the activities and the people until a permanent manager is found.

Key Implementation Challenges

The annual planning process provides an implementation architecture around which the alignment process can be executed, and each combination provides alchemy with different results. Following are the eight

Management-Level Staff Authority and Responsibilities

The following relates only to the people who report to the specific manager.

HR Comments

Authority/Responsibilities	ADG	Dept.	Section	Unit L3	Unit L4	Supervisor	Foreman
1. Coordinate activities	X	X	X	X	X	X	X
2. Make work assignments	X	X	X	X	X	X	X
3. Evaluate quality of work	X	X	X	X	X	X	X
4. Set performance objectives	X	X	X	X	X		
5. Evaluate staff performance				X	X	X	X
6. Select the staff	X	X	X	X	X		
7. Define who should be fired	X	X					
8. Approve promotions recommended by the managers that report to him or her							
9. Recommend staff for promotion	X	X	X	X	X		
10. Prepare educational requirements	X	X	X	X	X		
11. Decide on who is promoted or hired into the area of responsibility	X	X	X	X	X		
12. Authorize absenteeism	X	X	X	X	X	X	

Responsibility							
13. Authorize vacations	X	X	X	X	X		
14. Handle the employees' personal problems	X	X	X	X	X	X	
15. Interpret technical policies and procedures	X	X	X	X	X	X	X
16. Interpret personnel policies and procedures	X	X	X	X	X	X	X
17. Provide career guidance	X						
18. Recommend salary increases		X	X	X	X	X	
19. Prepare budgets	X	X	X	X	X	X	
20. Approve new projects under 1M AED	X		X				
21. Recommend new projects	X	X	X				
22. Give out rewards	X	X					
23. Communicate DM vision, values, and objectives	X	X	X	X	X	X	X
24. Improve the group's procedures	X	X	X	X	X	X	X
25. Motivate the group	X	X	X	X	X	X	X

FIGURE 6.13

Management-level authority and responsibilities matrix.

alignment implementation checkpoints for corporate, business units, and support units of a typical multibusiness organization to focus on and measure during the annual planning process.

1. Enterprise value proposition: The corporate office defines strategic guidelines to shape strategies at lower levels of the organization.
2. Board and shareholder alignment: The corporation's board of directors reviews, approves, and monitors the corporate strategy.
3. Corporate office to corporate support unit: The corporate strategy is translated into those corporate policies that will be administered by corporate support units.
4. Corporate office to business units: The corporate priorities are cascaded into business unit strategies.
5. Business units to support units: The strategic priorities of the business units are incorporated in the strategies of the functional support units.
6. Business units to customers: The priorities of the customer value proposition are communicated to targeted customers and reflected in specific customer feedback and measures.
7. Business support units to suppliers and other external partners: The shared priorities for suppliers, outsourcers, and alliance partners are reflected in business unit strategies.
8. Corporate support: The strategies of the local business support units reflect the priorities of the corporate support unit.

Using these eight implementation checkpoints as a point of reference, an organization can measure and manage the degree of alignment, and hence the synergy, being achieved across the organization.* Organizations that master this process can create competitive advantages that are difficult to dislodge. The insightful Spanish Jesuit Baltasar Gracian, almost 400 years

---

* Contained in some key concepts presented in this chapter were some generic definitions of planning. There are other definitions focusing more directly on the long-range planning process. For instance, according to David Hussey, "When a manager talks of corporate planning he is referring to a comprehensive business process which involves many types of planning activity. ... Corporate planning includes the setting of objectives, organizing the work, people, and system to enable those objectives to be attained, motivating through the planning process and through the plans, aligning all of the vectors, measuring performance and so controlling progress of the plan, and developing people through better decision-making, clearer objectives, more involvement, and awareness of progress." See David Hussey, *Corporate Planning: Theory and Practice* (New York: Pergamon, 1974, p.23) .

ago, captured the spirit of organizationally aligned Strategic Planning in words that fittingly sets the tone of this book:

> Think in anticipation: today for tomorrow, and indeed for many days. The greatest providence is to have forethought for what comes. What is provided for does not happen by chance, nor is the man who is prepared ever beset by emergencies. One must not, therefore, postpone consideration till the need arises. Consideration should go beforehand. You can, after careful reflection, act to prevent the most calamitous events. The pillow is a silent Sibyl, for to sleep over questions before they reach a climax is far better than lying awake over them afterward. Some act and think later, and they think more of excuses than consequences. Others think neither before nor after. The whole of life should be spent thinking about how to find the right course of action to follow. Thought and forethought give counsel, both on living and on achieving success.

Summary of Organizational Restructuring

One of the alignment basics to creating the future is the ability of the leaders to "engineer" or design the entire end-to-end value stream. Experienced managers know to redefine business designs and processes when implementing new forms of value. What is different in this new environment is the widespread implementation challenges involved with tactical synergistic clusters, business ecosystems, coalitions, cooperative networks, or sourcing to create end-to-end value streams. Tactical Business Communities (TBCs), as these networks of relationships are often known, are designed to align and to link businesses, customers, and suppliers in order to more uniquely create a new business organism and business model.

Microsoft illustrates the operational and tactical challenges involved and provides us with a glimpse of how to anticipate changing customer experiences by operationally realigning several value chains, including travel (Expedia), automotive sales (CarPoint), real estate (HomeAdvisor), and finance (Investor). The ultimate success of the new organizationally aligned value chains lies in the experimentation challenges and value they bring. This often depends on an integrated infrastructure that features the tactics of easy-to-use, sophisticated customer interaction. The targets of these online services are many and varied. According to Microsoft documents prepared as part of a 3-year planning process, Microsoft plans to win a major share of not only the $60+ billion advertising market, but also of sales distribution charges in the markets for airline tickets ($100+ billion), automobile sales ($330+ billion), and retail goods (over $1.2 trillion). Although this alignment initiative still remains unproven, Microsoft is a great example of a market leader that survived a competitive attack from Netscape and came out leaner, meaner, and stronger, while surviving the many implementation challenges involved.*

> Your present organization structure may be working for you but is there another one that will give better results?
>
> —HJH

---

* This strategy was originally outlined in a *Wall Street Journal* article titled "Microsoft Moves to Rule On-Line Sales" (*The Wall Street Journal*, June 5, 1997, sec. B, p. 1). Although Microsoft appears to have mastered the art of driving in turbulent weather, it's not very difficult to drive a car fast on a crowded freeway in good weather. But the worse the weather and heavier the traffic, the more frequently you have to change direction and speed. However, few organizations are capable thriving in demanding, changing of marketplaces that defy the normal business rules and forecasts.

7

Phase IV. Staffing

Your processes are important, but having the right caliber people using the processes will make or break the organization.

—HJH

Phase IV of the Organizational Alignment Cycle is defining the skills that the staff needs to have in the future-state organization and also defining how to acquire them (Figure 7.1).

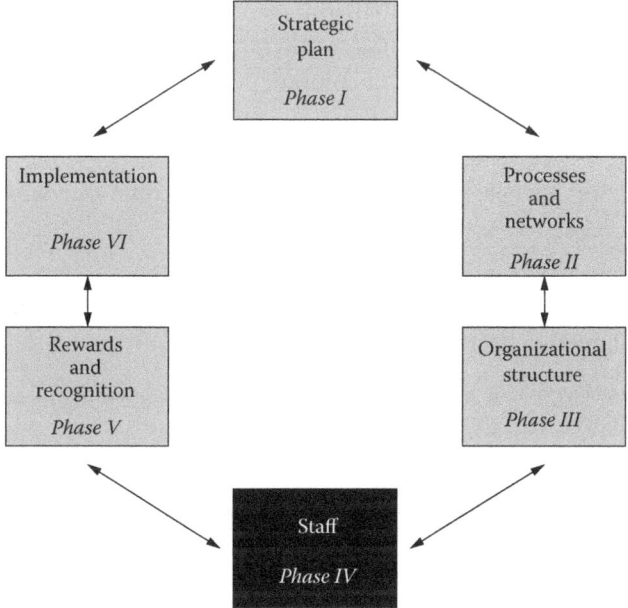

FIGURE 7.1

Phase IV of the Organizational Alignment Cycle.

INTRODUCTION

Now that the organizational structure and its supporting processes have been defined and documented, it is important to understand how well the people skills aligned with the processes that they will be operating and using.

To accomplish this, the organization should do a skill assessment of the present employees and then use the Strategic Plan to define the types of skills that will be needed in the future. Never mind what worked, or did not work, in the past. The rules have changed. It is not enough to define how many accountants, salespeople, development engineers, or assembly workers you have versus the quantity you will need 5 years from now. You need to define what skills they have or don't have that will be needed in the future.

- Will the salespeople need to have additional sales skills like experience accessing the Internet, creating PowerPoint presentations, and conducting virtual conferencing?
- Will we need fewer assembly workers, and will the ones who remain need to be highly skilled in computer operations?
- Will the marketing people and our service personnel need to speak Chinese and Russian?
- Will our purchasing personnel need to become our supply chain managers?

For sure, in most organizations, a much higher percentage of our personnel will need to become computer literate.

THE TWO APPROACHES TO SOLVING THE SKILL MIX PROBLEM

Now is the time to start to plan for the needed skills change. As an organization, you have two options:

1. Train your personnel.
2. Hire people to fill the skills void.

The correct answer for most organizations is a combination of the two. First, stop hiring people who don't fill a skill void, thereby letting attrition take care of the surpluses. If you need a person to do a job today that will result in a surplus skill, hire a temporary person or outsource it.

Alignment is slowly affecting the distribution channels through which consumers and businesses have traditionally bought and sold goods and services. The online channel provides sellers with the ability to reach a global audience. The new organizational structure and the changing business requirements frequently result in making our operating procedures and personnel practices obsolete and thus requiring them to change. In the previous phase (Phase III) the operations manual should have been already updated to reflect the vision, mission, new organizational structure, and new processes as well as the roles and responsibilities of the management team.

Often the personnel procedures, that is, career planning, dual ladder, appraisal systems, promotional requirements, and so forth, need to be changed. Often the measurement system needs to be updated to reflect the new requirements. This does not mean just to apply it to the corporate balanced scorecard but also to put measurements in all the way down to Individual Performance Indicators (IPIs), which apply to the natural work teams. These measurement changes are also reflected into the employees' performance objectives. In many cases the job descriptions need to be updated. For example, often the management's job description only reflects the technical side of their assignment. A good job description for a manager needs to cover all three parts of the assignment:

- Technical
- Business
- Personnel Management

Increasingly, end-to-end Organizational Alignment differentiates the winners from the rest of the pack. The implementation lesson learned is to begin with customer needs and work back along the fulfillment chain. This outside-in strategy requires thinking from the customer's perspective while working inward into the company's capabilities and direction. Microsoft's aligned operational infrastructure enables innovation and allows the firm to deliver newer and richer customer experiences. Using this Organizational Alignment strategy, Microsoft is attempting to create an entirely new set of service dynamics in various industries.

DEFINING SKILLS REQUIREMENTS

First we need to define what skills we are using today. There are three ways of doing this:

1. Ask the managers to record the information related to the skills that they are presently using.
2. Review the job descriptions and prepare a complete list of skills that are required based upon the job descriptions.
3. Perform a combination of 1 and 2.

We believe that approach 3 is the best one, that is, first review the job descriptions, and define what the job skill requirements are. Using this information, prepare a preliminary job skills list.

This preliminary job skills list is circulated to the management team so that they can update the list to reflect present skill requirements. Summing these inputs provides a good view of today's skill set.

Next you should review the Strategic Plan to define additional skills requirements that are needed. Typically these skills will be grouped in categories. The following is the typical example

- Performance Improvement
- Business and Financial
- Leadership
- Human Resource

Leadership is the principal driving force in behavior differentiation because companies, like all systems, tend to lose their edge without a constant infusion of "leadership energy."

—Terry Bacon, founder, Lore International Institute

Figure 7.2 is a sample of typical performance improvement skills.

In organizational development, performance improvement is the process whereby the organization looks to modify the current level of performance in order to achieve a better level of output. In aligning the organization structure with the Strategic Plan, performance measurement and improvement are important components that can enable the organization to know how much improvement the organizational change has achieved. Figure 7.3 is a list of typical business and financial skills.

Change Management	Communication
Continuous Improvement Management	Design of Experiments—Simulation
Measurement Analysis	Performance Evaluation
Poor Quality Cost	Process Improvement
Problem Solving	Process Redesign
Quality Reporting	Six Sigma Implementation
Team Building	Time Management
Teamwork	

FIGURE 7.2
Performance improvement skills.

Balance Scorecard Implementation	Financial Analysis
Benchmarking	Financial Management
Budgeting	Marketing
Investment Management	Outsourcing
ISO 14000	Project Management
ISO 9000	Project Portfolio Management
Privatization	

FIGURE 7.3
Business and financial skills.

> Distance reduces collaboration and so trust among members builds more slowly. This is especially problematic since the need for recognizing members' interdependencies may be high.
>
> **—Elaine Crowley, founder, The Crowley Group**

Business and financial strength are fundamental to the success of any organization. Based upon our experience, all Strategic Plans have new and increased requirements related to business and financial skills. Meeting the need of the next 5 to 10 years can only be achieved through sound business and financial management. Figure 7.4 is a list of typical leadership skills.

Leadership can be described as having the ability to lead, guide, direct, or influence people. In line with most Strategic Plans, the organization needs leaders with the leadership skills to successfully lead the transformation process. Such leaders usually have good problem-solving skills, are results oriented, are methodological in their approach, are organizationally savvy, are willing to accept responsibility for decisions, and are highly motivated. Figure 7.5 is a list of typical human resources skills.

Behavior Modification	Coaching
Creative Thinking	Empowerment
Delegation	Strategic Management
Meeting Management	Strategic Partnership Management
Motivating People	Succession Planning
Organization Development	Transparency
Building Trust	Knowledge Management

FIGURE 7.4
Leadership skills.

360-Degree Analysis and Feedback	Human Resources Management
Career Planning	Initiative
Competencies Analysis	Innovation
People Management	Rewards and Recognition
Personal Development	Self-Improvement and Self-Learning

FIGURE 7.5
Human resources skills.

Most organizations are in the process of transforming the Personnel Department into a Strategic Human Resources Department. This requires all the competencies and skills identified in Figures 7.4 and 7.5 and more, some of which will need to be developed for all levels of management. Figure 7.6 is a list of skill requirements that we generated for one of our government clients.

> Succession planning, especially in the top ranks of the organization, is taken seriously, but not urgently.
>
> **—Michael H. Shenkman, founder, Arch of Leadership**

OUTSOURCING

Often the 5-year plan will include outsourcing non-core activities to reduce cost and improve quality. The outsourcing activities often have a major impact on the skill mix that the organization will need. Activities like human resources, Accounting, Inventory Management and IT are very attractive outsourcing candidates. We have lately even seen Quality

TYPICAL SKILLS AND COMPETENCIES		
360-Degree Analysis and Feedback	Facilitation	Privatization
Accounting	Feasibility Studies	Problem Solving
Adaptability	Financial Analysis	Problem Tracking
Analytical Thinking	Financial Management	Process Control
Architectural Design & Planning	Forecasting	Process Improvement
Assets Management	Hazardous Waste Handling	Process Redesign
Balance Scorecard Implementation	Health Control	Project Management
Behavior Modification	Hospitality	Project Portfolio Management
Benchmarking	Human Resources Management	Public Relations
Best Practice Analysis	Information Management	Public Safety
Budget preparation	Information Security	Quality Reporting
Budgeting	Initiative	Real Estate Management
Building Inspection	Innovation	Recreational Management
Career Planning	Internet Operations	Resource Management
Change Management	Interpersonal	Rewards and Recognition
City Planning	Inventory Management	Risk Management
Coaching	Investment Management	Scheduling
Communicable Disease Control	ISO 14000	Self-Improvement and Self-Learning
Communication	ISO 9000	Sewage Control and Processing
Competencies Analysis	Knowledge Management	Six Sigma Implementation
Complaint Handling	KPI Development	Statistical Analysis
Computer Literacy	KPI Tracking	Stormwater Technology
Construction Management	Leadership	Strategic Management
Continuous Improvement Management	Legislation	Strategic Partnership Management
Contracting	Marketing	Strategic Planning
Coordination	Materials Standardization	Stress Management
Cost Benefits Analysis	Measurement Analysis	Succession Planning
Cost Control/Management	Media Relations	Supplier Certification
Creative Thinking	Meeting Management	Supplier Relations
Customer Needs Analysis	Motivating People	Supplies Audit
Customer Service	Negotiation	Supply Chain Management
Data Analysis	Organization Development	Team Building
Data Mining	Outsourcing	Teamwork
Decision Making	People Management	Testing/Inspection
Delegation	Performance Evaluation	Time Management
Design of Experiments—Simulation	Personal Development	Traffic Control/Management
Document Control	Planning	Transparency
Drainage Maintenance	Policy Development and Deployment	Trust
E-Government	Pollution Analysis	Vehicle Maintenance
Empowerment	Pollution Controls	Waste Management
E-Training	Poor Quality Cost	Work Load Analysis

FIGURE 7.6

List of skill requirements.

Assurance and Quality Control being outsourced. We suggest you take the processes that are planned for outsourcing and remove these skill requirements from your skill requirement list.

> Just as investors should diversify their investments, corporations should diversify their IT projects.
>
> **—Vin D'Amico, founder, Damicon, LLC**

PRESENT EMPLOYEES

Often an organization really doesn't know what skills they have on hand. For this reason we find it beneficial to have all the employees update the skills list in their personnel file at least every 2 years. Often people have been trained or educated in skills that they are not putting to use in the organization they are presently in. An engineer may have experience as a manager at another organization but has only been assigned to do product-engineer type work at the organization he or she is in. A truck driver may be very effective using a computer because he or she plays with it at home but has no opportunity to use it during the working hours. Updating these lists will provide you with an excellent understanding of where your strengths and weaknesses are.

MANAGEMENT SKILLS

If your organization is like most, its restructuring activity reduced the number of levels of management and the number of managers required. The question is "How do you get rid of these excess managers?" The answer is "We don't. Our managers are some of the best people we have. We made them managers and they have done outstanding work and we do not want to lose them."

The real problem is "What do we do with these surplus managers?" We like to start by developing a dual ladder that provides the excellent, technical/professional people with the same rewards and recognition that the various levels of management have. For example, there could be five levels of codes for engineers. The following are typical engineering levels and the equivalent management level.

1. Junior Engineer—No equivalent management level
2. Engineer—Equivalent management level; foreperson/team leader
3. Senior Engineer—Equivalent management level; manager
4. Project Engineer—Equivalent management level; project manager (middle manager)
5. Fellow—Equivalent management level; vice president

Conduct a career profile analysis to define the part of the job from which individual managers get the most satisfaction. It will also help to design which area these managers will reach the highest level in. Excellent workers often get assigned as managers, and they take on this new position because it's their only growth path. Some new managers may also take on this new position primarily because it gives them more hands and legs to get the job done, and they don't realize that they have taken a very different growth path, which may or may not be suitable for them.

The career profile analysis is a 1-week program during which the abilities and weaknesses of participants are defined and career paths are developed for them. Here again we find that many of the managers decide to stay out of management if there is another equally good opportunity.

Usually when individuals choose to step out of management, they are looked upon as failing. So although they may not like doing a manager's job—holding appraisals, listening to family problems, and chastising someone on the team because that person comes into work 10 minutes or more late—they decide to keep the management position even though they don't enjoy it. As a result, these managers focus on the technical side of the job and let the people side slip. Good managers have to get their kicks out of promoting people, not doing technical work.

We know of one company whose sales fell off, causing them to have almost 40% surplus of managers. In place of letting these excess managers go, the company started a program called the Technical Vitality Program. In this program one third of the managers were assigned to a technical job for 1 year and at the end of the year they were brought back into management. In this case over 25% of these managers decided not to come back into management because they enjoyed the technical area better. The managers that left management were not looked upon as failing because it was well known that this decision was their personal choice; they could choose to stay in management or not.

SUMMARY

Because the improvement processes and Strategic Planning activities usually reduce the number of people required to operate the processes, job security has become one of the most critical and complex political and economic issues facing top management today. How can you expect your employees to give freely of their ideas to increase productivity and minimize waste if it means that their job or a friend's job will be eliminated? If you start a continuous improvement process and then have layoffs, what you are going to end up with is a continuous sabotage process.

> Nothing is without trade-offs, and offshoring is not an exception. Before sending software development offshore, it is important to consider all of the potential drawbacks.

—David Vogel and Jill Connelly, *Handbook of Business Strategy*, 2005

Large layoffs produce sudden, substantial stock gains. These gains occur because the impact of the removed employees has not reached the customer and the compensation has been removed from the bottom line, making the organization appear to be more profitable than it really is. But in the long run, the downsizing has a negative impact. CEO Frank Poppoff of Dow Chemical put it in this way, "Layoffs are horribly expensive and destructive of shareholder value." This is because most organizations downsize rather than "right size." What organizations have been doing is cutting "X" percent per area—not eliminating any work, just distributing the same amount of work to fewer people. This is an insult to your employees because it is like saying that management thinks they have not been doing a fair day's work for a fair day's pay. There are at least 16 other ways to handle the problem of surplus employees that should be looked at before the organization lays off anyone.* The cost to lay off and replace personnel is growing all the time. Dow Chemical estimates its costs from $30,000 to $100,000 for technical and managerial type personnel. Layoffs not only cost the organization money and some of its best people, but when it comes time to hire, the best people will not trust this type of organization and will not come to work for them.

---

* For more information see H. James Harrington, *Total Improvement Management* (New York: McGraw-Hill, 1995).

An alternate approach of a golden parachute or early retirement is equally bad. The people who leave are all the best performers who will not have a problem finding new jobs. The deadwood, who barely meet minimum performance, stay because they know it will be hard to find an equally good job in today's job market.

Employees can understand that the organization needs to cut back when demand for the product falls off, and they can accept that. The problem we face is what happens to the employees whose jobs have been eliminated as a result of organizational realignment. We know that organizational realignment is designed to improve productivity. But if our share of the market does not keep pace with our productivity improvement, what will management do with the surplus employees? To cover this scenario and to alleviate employees' fears, top management should release a "no-layoff policy." A typical no-layoff policy would state: *No employee will be laid off because of improvements made as a result of the performance improvements. People whose jobs are eliminated will be retrained for an equivalent or more responsible job. This does not mean that it may not be necessary to lay off employees because of a business downturn.*

You will note that a typical no-layoff policy does not guarantee that employees will not be laid off as a result of a downturn in your business. It only protects employees from being laid off as a result of the improvement process. These are the people who would still be working if the improvement process had not been implemented. Federal Express Corporation has a no-layoff philosophy. Its "guaranteed fair treatment procedure" for handling employee grievances is a model for organizations around the world.

We know of one organization that was able to eliminate about 210 jobs as a result of its improvement process. As they started the improvement process, they put a freeze on new hires and used temporary employees to cover the workload peaks. This was reviewed with the labor union leaders and they concurred with the use of temporary employees to protect regular employees' jobs. As a result, attrition took care of about 145 surplus jobs. The organization then held a contest to select 65 employees, who were sent to a local university to work toward an engineering degree. While at school, these employees received full pay, and their additional expenses were paid for by the organization. At the end of a year, all of them were back working on better jobs than they had before and most of them went on their own to get their degrees. But the real miracle took place inside the organization. The word got out that if your job was eliminated, you would not be laid off but sent to school. All of a sudden everyone was looking for ways to eliminate their jobs so they could go to school.

Increasingly, end-to-end Organizational Alignment differentiates the winners from the rest of the pack. The implementation lesson learned is to begin with customer needs and work back along the fulfillment chain. This outside-in strategy requires thinking from the customer's perspective while working inward into the organization's capabilities and direction. Microsoft's aligned operational infrastructure enables more innovation. This allows Microsoft to deliver newer and richer products to its customers.

Our employees are the stuff of life for our business.

—HJH

8

Phase V. Rewards and Recognition System Design

People want to be praised and appreciated even more than they want money or sex.

—HJH

Phase V of the Organizational Alignment Cycle is developing a set of reward and recognition's processes that reinforce the organization's Strategic Plan (Figure 8.1).

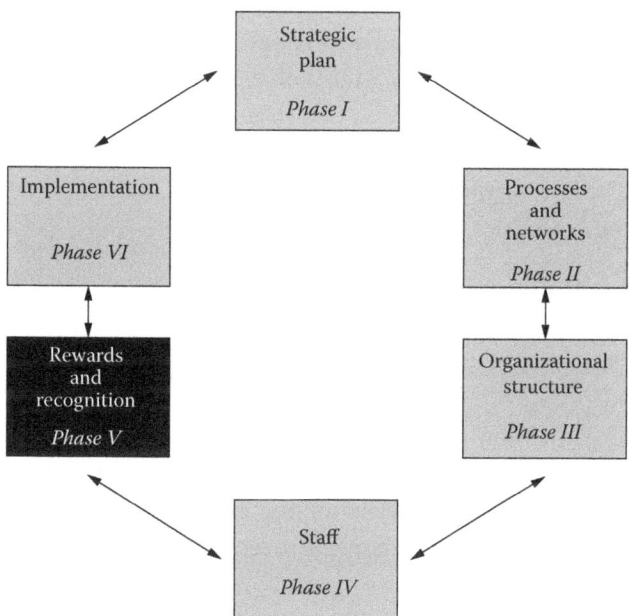

FIGURE 8.1
Phase V of the Organizational Alignment Cycle.

INTRODUCTION TO REWARDS AND RECOGNITION (R&R) PHASE

The purpose of the reward and recognition system is to reinforce behavior patterns that support the organization's mission, vision, and Strategic Plan. It is important to remember that the R&R system is designed to pay for performance, whether it is physical or mental, and to recognize people who have achieved something over and above what they are normally expected to accomplish. The individual's salary is the reward for a normal job well done. Only the truly outstanding people should receive contribution awards and special recognition.

To determine the current situation, the following questions should be asked of management:

1. Do people perceive rewards and recognition for just meeting goals that are not stretch goals?
2. Are rewards based on what the tracking system captures and reports?
3. Does the reward system recognize short- and long-term achievements?
4. Do the elements of the pay system (merit increases, bonuses, extra compensation) support the Strategic Plan and the organization's goals?
5. Are employees recognized and rewarded for initiatives and actions in the best interests of long-term objectives and strategies?
6. Are promotions perceived as tied to achievement of plans and goals?
7. Are merit increases tied to activities and measures that reflect a unit's plans and priorities?
8. Do you have team awards?

As many as 25% of good employees, who quit their jobs, cite a lack of appreciation as their reason.

—Survey by Robert Half International, Inc.

As the organization changes, the R&R system, as well as the minimum acceptable performance standards, need to change to reflect the new required behavioral patterns for the executives and the employees. Figure 8.2 is the Harrington Change Process Chart.

- P1 stands for the present performance standard. You will notice there is a wide range of variation from the top or best performance to the bottom or poorest performance, which is unacceptable.

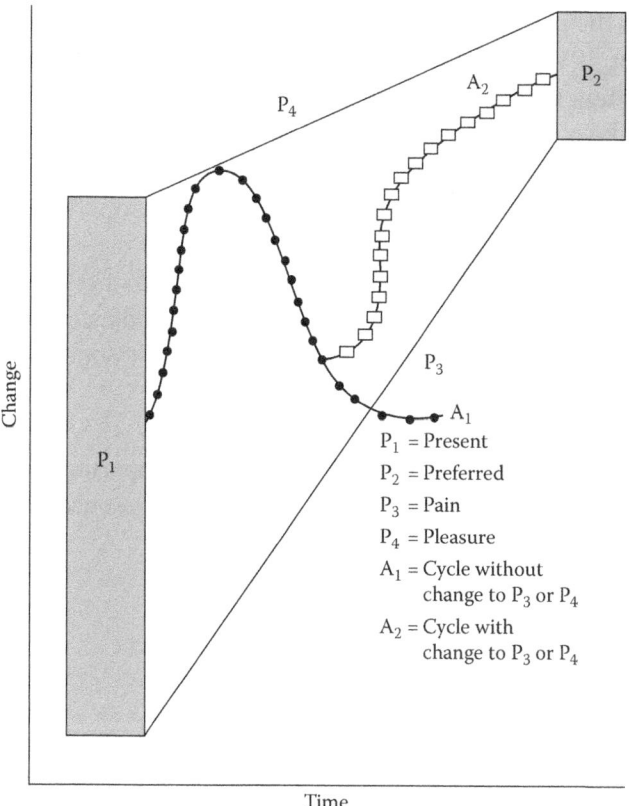

FIGURE 8.2
Harrington's change process chart.

- P2 stands for the performance required to meet the new strategies and cultural requirements of the future state. You will note that the variation is small, and the minimum required performance is even better than the best performance in the present state conditions. The dotted line A1 indicates what happens when an improvement project is implemented and the performance standards don't change. There is usually an improvement when the focus is on the improvement process, but after the project is completed and management's attention is diverted to another area, the performance settles back to where it was.
- P3 stands for the pain level; it is the point where the employee/manager is subject to pressure to improve and is often removed from the assignment if there is no improvement.

- P4 is the pleasure point. This is the point where the individual or the group is recognized, rewarded, or both, for performing in an outstanding manner. This can be as simple as a pat on the back and a thank-you from the manager to something as important as a promotion. You will note that in Figure 8.2, people are subject to pain in the future state solution (P2) that they would have received pleasure for in the (P1) present state process.
- A2 shows what happens to the average person's performance when their pain (P3) and pleasure levels (P4) are adjusted to be in line with the desired performance after the improvement process has been completed.

It would be a big mistake to fail to point out the importance of a rewards and recognition system for excellence in discussing resource management because it's such an important human enabler.

KEY OBJECTIVES

Everyone hears "thank you" in a different way.

—HJH

A good recognition process has six major outcomes and objectives:

1. To provide recognition to employees who make unusual contributions to the organization to stimulate additional effort for further improvement.
2. To show the organization's appreciation for superior performance.
3. To ensure maximum benefits from the recognition process by an effective communication system that highlights the individuals who were recognized.
4. To provide many ways to recognize employees for their efforts and stimulate management creativity in the recognition process. Management must understand that variation enhances the impact.
5. To improve morale through the proper use of recognition.
6. To reinforce behavioral patterns that management would like to see continued.

There are as many ways to recognize people as there are people to be recognized.

—Eric Harvey, author

Why does recognition matter? George Blomgren, president of Blomgren Consulting Services, Ltd., puts it this way, "Recognition lets people see themselves in a winning identity role. There's a universal need for recognition and most people are starved for it." A National Science Foundation study made the same point. "The key to having workers who are both satisfied and productive is motivation, that is, arousing and maintaining the will to work effectively—having workers who are productive not because they are coerced but because they are committed."[*]

We're trying to get "the ideal deal"—the combination of rewards and processes that create the utmost employee commitment and engagement.

—Daryl David, Human Resources Director, Xilinx

SEVEN MAJOR TYPES OF R&R

The seven major types of rewards and recognition are as follows:

1. Financial compensation
2. Monetary awards
3. Personal public recognition
4. Group public recognition
5. Private recognition
6. Peer recognition
7. Organizational awards

Financial Compensation

A study by the Public Agenda Foundation[†] revealed the following:

- Employees are not rewarded for putting out extra effort.
- Almost two thirds of the employees would like to see a better connection between performance and pay.

---

[*] National Science Foundation Study on Motivation, NSB-09-2, Investing in the Future, January, 2009, pg. 3. Draft for Public Comment.
[†] H. James Harrington's Technical Report 97.002, Total Improvement Management (2007).

- More than 70% of employees think that the reason work effort has deteriorated is because there is no connection between pay and performance.

Compensation is what you give people for doing the job they were hired to do. Recognition, on the other hand, celebrates an effort beyond the call of duty.

—"How to Profit from Merchandise Incentives"*

In most organizations it is generally accepted that the best performers are the ones who produce the most, on time. They often are the ones who come in on Saturdays and work late most nights while the ones that come in late and leave early are considered poor performers. For a long time, management believed that the poor performers would look up to the best performers and want to copy them. The truth of the matter is that the best performers look at the poor performers and question if the slight difference in pay is worth the added effort. As a result, the best performers start to slack off, dropping down closer to the average performer level. The average performer sees that the best performers are cutting back and they think it is okay for them to cut back, doing less and producing poor work. This is a vicious cycle that leads to a downward spiral. So we pour more money into technology to make up for it just to hold our own. People who do twice the work should get twice the rewards—that is, pay for performance.

Bell Atlantic is trying a new pay for performance system. It is holding back pay from its 23,000 manager-level employees and will distribute this money to them at the end of the year, with more of it going to the high-performing managers. Salaries are important, and the means for relating quality and productivity to salary and an effective performance appraisal system are important ingredients for motivation. But other types of financial compensation can also motivate improvements in productivity and quality.

In addition to salaries, typical financial compensations are

- Commissions
- Piecework pay

---

* See *Where Americans and Business Leaders Agree/Disagree on Business Ethics,* 2004, Public Agenda Corporation, a nonpartisan opinion research organization helping Americans explore and understand critical issues. Public Agenda was founded by social scientist and author Daniel Yankelovich and former Secretary of State Cyrus Vance in 1975. This study also included jobs, greed, and weak values as real concerns; sharp differences in what citizens and business leaders think about layoffs; and what makes a good or bad corporation?

- Employee stock plans
- Cash bonuses
- Gain sharing

Challenges with Commissions

Have you ever noticed how enthusiastic your Avon salesperson is when he or she knocks on your door? Did you see the same degree of enthusiasm in your organization's order clerk the last time you placed an order for a gross of paper clips? The difference is that the clerk in the order department is working for a salary, whereas the door-to-door salesperson is working for a commission. The better the service the door-to-door salesperson provides the customers, the more return sales are generated and the more money is earned.

The successful salesperson is always turned on because he or she looks at the last sale as additional compensation that wasn't there before finding the client. But the clerk in the order department looks at the new order of paper clips as additional work that wasn't really needed and that has no impact upon his or her personal financial status.

Customers in the service industry are being turned off today, not by price but by carelessness, discourtesy, and disinterest. Customers want a caring and friendly salesclerk, a cheerful "Good morning, may I help you?" They want to be treated as valuable individuals who are important to the clerk and to the organization. If you show them this type of consideration, they will buy more and come back again and again. Commissions provide a means to motivate certain employees to do more and better work. Commissions work to improve the salesperson's productivity. This same principle can be applied to other white-collar activities. For example, a design engineer could receive a percentage of the profit from a product he or she designed.

Challenges with Piecework Pay

Piecework was a popular method of increasing productivity in the first part of the 20th century and still is widely practiced in some parts of the world. In essence, this system pays the employee a share of the value added to the item being processed, based upon the effort, skill, and time required to complete the task. Along with the sweatshop, piecework has slowly been phased out of the American scene because although it did increase raw productivity, it did not emphasize quality and it resulted in a great deal of suboptimization.

Challenges with Employee Stock Plans

Employee stock plans are increasingly popular because they provide an effective means of focusing the employees' attention on the business aspects of the organization by allowing them to share in the profits. They also help to break down the "we and they" feeling many employees have about the organization, because from their standpoint the stockholders are the ones who make all the money from the employees' hard work.

In commenting about the stock purchase plan Hewlett-Packard (HP) uses, John Young, its former president, said, "The company contributes $1 for every $3 an employee puts in. More than three-fourths of our people participate in the plan. The result has been that since employees own part of the company, they feel ownership for some of the company's problems and successes."*

More than 10 million American workers already belong to employee stock plans, and within 15 years more than 25% of all American workers will join the parade. The number of new employee stock plans is growing at about 10% per year, spurred on by the 1984 U.S. tax incentive. A recent study of 360 high-technology companies conducted by the National Center for Employee Ownership concluded that companies that share ownership with their employees grew two to four times faster than companies that did not have employee stock plans.†

The object of an employee stock plan is literally to make capitalists out of all employees. Many plans have been developed and used to directly tie the employees' economic future into the success of the organization. One of the more popular plans now in use is called the employee stock ownership

---

* A 2001 interview with John Young on the subject of employee compensation and stock ownership. Although he retired in 1991, John Young remained an active spokesperson for employee involvement and stock ownership as a source of motivation and employee pride.

† National Center for Employee Ownership. In 2007, the largest and most significant study to date of the performance of employee stock ownership plans (ESOPs) in closely held companies was conducted. Douglas Kruse and Joseph Blasi of Rutgers found that ESOPs appear to increase sales, employment, and sales per employee by about 2.3% to 2.4% per year over what would have been expected absent an ESOP. ESOP companies are also somewhat more likely to still be in business several years later. Moreover, ESOP companies are substantially more likely to have other retirement-oriented benefit plans than comparable non-ESOP companies. While the results are in line with previous studies, no study of closely held companies yet has matched the scope of this one.

plan. Lewis O. Kelso, a San Francisco lawyer, developed the theory behind this kind of plan.

Challenges with Cash Bonuses and Gain Sharing

> You need to go after and reward the short-term wins while maintaining a clear sense of direction.
>
> **—Joe W. Forehand, CEO, Accenture**

HRD Training Journal reports that more than 75% of manufacturing companies in the United States have an executive bonus plan.[*] A study of 1000 organizations revealed that organizations with executive bonus plans average better than 40% pretax profit. Cash bonuses and gain sharing are not new—they can be traced back to Roman civilization. In modern times, they have been a proven effective way for an organization to share its profits with its employees. Suggestion awards should be separate from the bonus system and should be paid directly to the employee who makes the suggestion.

Organizations that use gain-sharing programs find their employees starting to think differently, starting to learn and use new vocabulary— words such as "profits," "gross sales," and "production costs" start to slip into their conversation because they see for the first time a direct correlation between their well-being and the well-being of the organization.

At General Motors, managers tell employees about the plant's direct labor costs, scrap and rework costs, and profits compared to targets that the organization has set for itself. This is information that only top management had in the 1970s. General Motors believes that providing employees with this kind of information helps to close the gap between management and labor. It has proven to be an effective way of aligning organization and employee goals and developing a partnership between the two groups that until recently were in opposition to each other.

In an international study on productivity conducted by Louis Harris and Amitai Etzioni, nearly two-thirds of the employees surveyed indicated that most employees would be happy to have their salaries linked to higher productivity. To take advantage of this opportunity for employee involvement through the years, different kinds of bonus and gain-sharing

---

[*] H. James Harrington's Technical Report 97.002, Total Improvement Management (2007).

programs have been developed. Most of them focus on an equal division of a pool of money among the employees, tied to their base salaries. Of these programs, the best known are the following:

- Scanlon plan
- Rucker plan
- Improshare plan

A government study of 36 organizations that are using productivity-sharing plans (17 Scanlon, 11 Improshare, and 8 Rucker) revealed the workforce savings averaged 17.3% for organizations with less than $10 million annual sales and 16.4% for organizations with more than $100 million in annual sales.*

Individual incentive or bonus programs are difficult (but not impossible) to administer in non-sales activities in large corporations. In small organizations, individual incentive programs have definite advantages. They require a lot of management attention and emphasis, but the resulting improvement is well worth the effort. Delta Business Systems, a $32 million company with headquarters in Orlando, Florida, applied 13 different incentive programs to its more than 2000 non-sales positions. For example:

- Warehouse workers divide up to $400 every 2 months for filling out orders on schedule, processing paper so that the organization receives the allowed cash discount, and keeping the operation working smoothly.
- For retaining their customer base, giving sales personnel leads, and getting maintenance agreements renewed, the field service technicians can increase their salaries by 3% to 25%.
- People in accounts payable were offered up to $200 a quarter to reduce outstanding unpaid bills. As a result, the long-term accounts payable were reduced by 50%.

Bryan King, president of the organization, put it this way: "If we can see how fast someone's canoe moves in the water, we provide an incentive for him to improve."

---

* H. James Harrington, *The Improvement Process* (New York: McGraw-Hill, 1987).

Monetary Awards

Monetary awards are another class of recognition. The word "award" indicates it is a unique recognition of an individual or small group for unusual contributions to the organization's goals. Monetary awards are one-time bonuses paid to the recipient immediately after an unusual or far-exceeds-expectations contribution. They may also be given to individuals for long-term, continuous, and high-level performance or unique leadership. The award should be specific, and management and fellow employees should perceive the person or people who receive the award as "special." The amount of the monetary award should vary based upon the magnitude of the contribution.

The many types of monetary awards reflect differing degrees of contribution. These three are typical:

- Suggestions awards—These types of awards tap the creativity of the employees. Best practices result in each employee turning in two suggestions per month with 80% of them being implemented. Examples of organizations that are performing at that level and how to accomplish it can be found in the book, *The Idea Generator*, by Norman Bodek and Buniji Tozawa, published by PCS Press in 2001.
- Patent awards—This type of award may present a problem to management. In most cases, the people who are applying for patents are being paid to develop them. Nevertheless, you want to encourage the individual who is generating new creative ideas that are generating hundreds of jobs and large revenues for the organization. Some organizations use a plateau award system. In these systems, the employee accumulates points based on the number of patents received and the potential contribution that each patent makes to the organization. As the employee moves up from one award level to the next level, the organization gives more meaningful awards.
- Contribution awards—An effective contribution award system must provide flexibility to management and equity to all employees. It must be based on actual contributions of the person and must be administered in all areas using the same ground rules. This discussion will refer to the contribution award system shown in Figure 8.3.

As the contribution to the organization increases, the value of the monetary reward increases, and it becomes increasingly difficult to obtain

Award Name	Dollar Amount
Outstanding contribution award	$5,000 to $200,000
Recognition award	$1,000 to $10,000
Weekend on the town award	$1,000
A night out for two award	$300

FIGURE 8.3
Contribution award system.

management approval to give the award. For example, a line manager should be able to give one of his or her employees a night out for two when the occasion warrants it. On the other hand, the outstanding contribution award should be supported with a detailed written description of the contribution and its impact on the organization. An organization recognition review board should review it to ensure that the award system is being interpreted equitably in all areas of the organization. The award should be presented at a formal meeting with the total function in attendance. In addition to the money, the employee should receive some special award jewelry (tie tack, ring, pin, etc.) and a framed certificate. The jewelry provides a continuous reminder to people that the program is viable and available to them. The money is soon spent and forgotten, but a Super Bowl ring lasts forever as a reminder of excellence.

Personal Public Recognition

An almost endless list of types of recognition does not directly involve money. Here are some ideas that should stimulate your thinking and give you something to build on:

- Promotions
- Office layout, size, or possibly view
- Trips to customer locations
- Organization recognition meetings
- Annual improvement conferences
- Jewelry
- Special parking spaces
- Articles in newsletter
- Public notice posted on the bulletin board (one plant in New York City has a huge billboard on top of the plant that flashes the names and accomplishments of special individuals)

- Employee's picture on a poster
- Verbal recognition at a department, division, or company meeting
- Special job assignments
- Plaque presented in front of fellow workers
- Plaque in the organization's entranceway with the employee's picture and name

Bob Nelson's book, *1001 Ways to Reward Employees* (published by Workman Publishing, New York City) provides many more good ideas.

Stacoswitch Corporation in Costa Mesa, California, provided all its supervisors with badges that read: "We do it right or we don't do it at all." Supervisors with the lowest reject rate had gold stars on their badges.

National Car Rental uses commemorative plaques with individual employee names that hang in its business offices around the United States. A typical plaque would read: "Carrie Harrington deserves national attention for outstanding performance, second quarter, 2009."

Jim McCann, CEO of 1-800-Flowers, sends people out to speak on behalf of the organization. He states, "Look for venues where your employees can spread the word about the organization and their enthusiasm.*

Group Public Recognition

Recognition of the group makes the group think that it's a winner and the members of the group get a sense of belonging that leads to increased organizational loyalty. Again, management has an unlimited number of ways to recognize a group for contribution. Typical ways include the following:

- Articles about the group's improvement in the organization's newsletter, accompanied by photographs of the group
- Department luncheons to recognize specific accomplishments
- Family recognition picnics
- Progress presentations to upper management
- Luncheons with upper management
- Group attendance at technical conferences
- Cake and coffee at a group meeting, paid for by the organization
- Department improvement plaques

---

* Eric Harvey, *180 Ways to Walk the Recognition Talk* (The Performance Systems Corporation, 2000, Dallas, Texas).

- Top management's attending group meetings to say thanks for a job well done
- Group mementos (pen sets, calculators, product models, etc.)

I can live two months on a good compliment.

—**Mark Twain**

Private Recognition

Of all recognition categories, private recognition is one of the most important because it directly relates to the interaction between management and the employee. The one-on-one interaction is very important in stimulating improvement and keeping morale high.

Many managers feel strange telling employees they are doing a good job, and frequently the employee has a hard time accepting compliments and reacts with comments such as: "Ah, knock off that bull!" or "Don't give me compliments, give me money!" But such comments don't mean they don't need the manager's appreciation. So don't let it prevent you from expressing your honest appreciation for a task well done. Employees need encouragement and must have management appreciation reinforce their good acts. A sincere pat on the back at the right time is much better and more effective than a swift kick in the pants at any time.

Typical ways that management provides private recognition to an employee are listed here:

- A simple, honest thank-you for a job well done, given immediately after the task is completed
- A letter sent to the employee's house by his manager or upper manager, thanking him or her for a specific contribution
- Personal notes on letter or reports, complimenting the originator on content or layout
- Sending birthday cards and work anniversary cards to an employee's house, thanking the employee for the contributions that were made during the past year, not with general statements but with specific examples that let the employee know that management knows that the employee is there and knows what he or she is doing

The performance evaluation that takes place every 3 months is an ideal time to give private feedback to the employee about accomplishments.

It shouldn't be the first time you have expressed your appreciation, but you should use it to reinforce the favorable work patterns and summarize employee accomplishments. The most basic rule of performance evaluations is "no surprises."

Peer Recognition

One of the most significant rewards is the sincere appreciation from your peers for a job well done. Frequently, these types of rewards are associated with professional societies, where the practicing professionals select the brightest and best in the profession to honor at an awards luncheon. Today, the same concept has become a positive motivating force within many organizations. These award recipients are selected by the employees, not by the management.

Often, management establishes the basic ground rules and financial constraints related to peer recognition rewards. Representatives of the employees should then define what types of behavior should be recognized and establish how these behaviors will be rewarded. Empowering the employees to select the reward frequently provides management with pleasant surprises related to the creative way employees apply the limited reward budget. Our experience indicates that employees usually plan something that is fun, and the winners are usually very touched. Typically when management prepares a rewards ceremony, it turns out to be a ceremony. When employees plan the same event, it turns out to be a party, and the gifts that are presented turn out to be from the heart rather than the pocketbook.

The importance of building organizational trust and pride in our employees' minds is a major objective of most improvement processes. Unfortunately, organizational pride has slipped in most Western organizations over the past 40 years. We need to start to rebuild the pride our employees had in our organizations, for when individuals start to take pride in the organizations they work for, they take more pride in the things they do because they do not want to tarnish the organization's reputation. One of the chief advantages that Japan, Inc., has over Western countries is the pride and dedication their employees have toward their organizations.

Organizational Awards

An effective way of gauging your employees' pride in your organization is to have the organization recognized by its peers as being outstanding. This is what the recent series of organizational improvement awards are accomplishing, as well as setting benchmarks for other organizations. Today, there are many award programs implemented throughout the world to recognize excellence in individual organizations. Some of them are:

- Deming Prize – Japan
- Japan Quality Control Prize – Japan
- Malcolm Baldrige National Quality Award – United States
- Shingo Prize – United States
- NASA Award – United States
- President's (Federal Government) Award – United States
- European Quality Award – Europe
- Australian Quality Award – Australia
- Best Hardware Laboratory – IBM – United States
- International Asia Pacific Quality Award – Asian/Pacific areas

Don't go around saying the world owes you a living. The world owes you nothing; it was here first.

—Mark Twain, author and humorist

IMPLEMENTING THE R&R SYSTEM

To develop an effective reward process, you must take many factors into consideration. These activities will help you avoid most reward process problems:

- Reward Fund—The organization should set aside a specific amount of money to use in the reward process. This amount will set the boundaries within which the reward process will operate.
- Reward Task Team (RTT)—This team will design and update the reward process.
- Present Reward Process—The RTT should pull together a list of all the formal and informal rewards that are now used within the organization.
- Desired Behaviors—The RTT should prepare a list of desired behaviors.
- Present Reward Process Analysis—It should review the present reward process to identify the rewards that are not in keeping

with the organization's present and projected future culture and visions.

- Desired Behavior Analysis—It now compares each desired behavior to the reward categories to see which category or categories should be used to reinforce the desired behavior. Each behavior should have at least two ways of rewarding people who practice the behavior.
- Reward Usage Guide—After defining the reward process, the RTT should prepare a reward usage guide. This guide should define the purpose of each of the reward categories and the procedures that will be used to formally process the reward. This guide will help management and employees understand the reward process and will help standardize the way rewards are used throughout the organization.

Why is it that we train our managers on how to handle poor performers, but never train them of how to say "thank you"?

—HJH

- Management Training—Management should be trained how to use the reward system, one of the most neglected parts of most management training processes. As a result, most managers are far too conservative with their approach to rewards, while others misuse them.

In creating a reward process, consider the following:

- Always have it reinforce desired behaviors.
- Reward for exceptional customer service and performance.
- Publish why rewards are given.
- Create a point system that can be used to recognize teams and individuals for small and large contributions. The employee should be able to accumulate points through time to receive a higher level reward.
- Structure the reward process so that 40% of the employees will receive at least a first-level reward each year.
- Structure the reward process so that the managers can exercise their creativity and personal knowledge of the recipient in selecting the reward.
- Provide ways that anyone can recognize a person for his or her contributions.
- Provide an instant reward mechanism.

SUMMARY

A soldier will fight long and hard for a bit of colored ribbon.

—**Napoleon Bonaparte**

No matter how much R&R you are using, there is always more that is needed. The following checklist will help you evaluate your R&R system.

- Be sincere. False praise is worse than no praise. The person that receives it knows he or she didn't deserve it and it makes the rest of the employees feel left out.
- Do it now. Give the R&R as soon as the desired behavior is recognized. The best time to reinforce a behavior is while it is being done and can be remembered.
- Pinpoint the individual. Be specific about why the individual is being rewarded. Words like "good job" have some effect, but it means a lot more if you say, "The graphs made it very meaningful to me. Keep up the good work."
- Vary the approach. Learn what turns each person on and vary the R&R to meet the individual's hot button.
- Select the award based upon the achievement. Things that support the business plan need bigger R&R than ones that don't. Give big awards to ones that have a big impact on the organization. Give many small ones often to reinforce culture changes.

Remember the six Golden Rules to build morale using R&R:

1. Train your managers to use the R&R system.
2. Personally congratulate employees as they are observed doing a good job.
3. Publicly recognize outstanding performers.
4. Send personal notes or e-mails to reinforce good behaviors.
5. Promote people based upon outstanding performance.
6. Organize meetings to celebrate the successes and to give out rewards.

Outstanding leaders go out of their way to boost the self-esteem of their personnel. If people believe in themselves, it's amazing what they can accomplish.

—**Sam Walton, CEO of WalMart**

9

Phase VI. Implementation

Even the best ideas and plans are not effective if they are implemented poorly.

—HJH

The first five phases of the Organizational Alignment Cycle have set the stage for the organization to successfully implement the changes with minimum risk of not meeting the organization's objective (Figure 9.1).

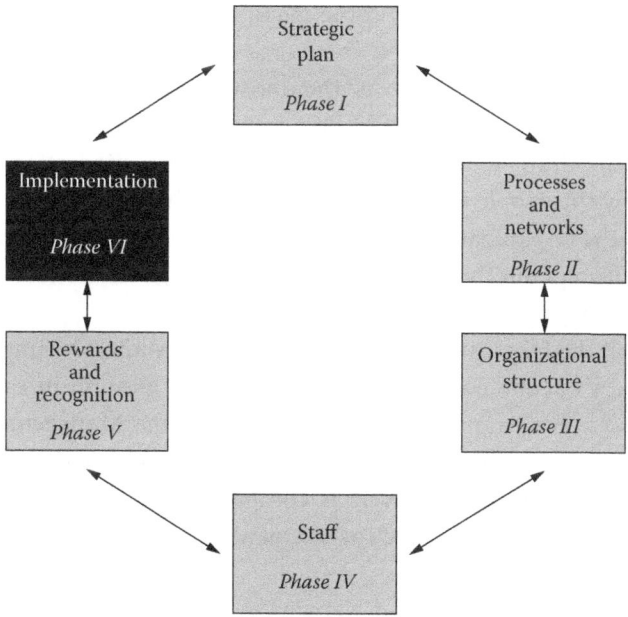

FIGURE 9.1
Phase VI of the Organizational Alignment Cycle.

INTRODUCTION TO IMPLEMENTATION PHASE

Let's look at where we are now. During Phase I, Strategic Planning, we developed or updated the organization's mission, vision, objectives, goals, strategies, and priorities for the short- and long-range activities. In Phase II, Processes and Networks, we redesigned the major processes and networks so that they were in line with the Strategic Plan. In Phase III, Organizational Structure, we restructured the organization to support the redesigned processes and networks. This restructuring also took into consideration the advantages and disadvantages of the different types of organizational structures and principles. In Phase IV, Staffing, we evaluated the skills and talents of our people and compared them to the future requirements as driven by the Strategic Plan. This gap analysis will be used to direct the training process for your people and your hiring practices. In Phase V, Rewards and Recognition, we redesigned the rewards and recognition systems to reinforce the new desired behaviors that our organization of the future will require to support its new culture. These activities often result in some major changes that impact the total organization.

Phase VI, Implementation, is the most critical phase of the total Organizational Alignment Cycle because all of the previous effort can go down the drain if the changes are not implemented effectively and accepted by the people who are affected by the changes. It is at this point that all of the Organizational Change Management efforts you have been applying throughout the first five phases of the Organizational Alignment Cycle really pay big dividends.

Although the Organizational Alignment Cycle shows the implementation Phase VI at the end of the cycle, in truth it goes on in each of the other phases. For example, Organizational Change Management activities start early in Phase I (Strategic Planning) and continue until the Organizational Alignment Cycle is complete. Also during Phase I the communication and employee involvement activities start and continue through the entire cycle.

During Phase II (Process and Network Analysis), the process upgrades are implemented in each process as it is redesigned. In Phase III (Organizational Structure Analysis) the organization's structure is changed as soon as a better one is defined. During Phase IV (Staffing) the staffing procedures are implemented and new training plans are put in place immediately. Surplus managers are cycled through a technical vitality program to put them into meaningful assignments. During Phase IV (Rewards and Recognition) reward procedures are redefined and the supporting training and documentation are released.

So as you can see, implementation is an integral part of all of the other five phases. As a result, we are going to present some of the approaches that help to implement the required changes in each of the five phases. We will present the following:

- Organizational Change Management
- Communication Planning
- Area Activity Analysis
- Business Process Management
- Outsourcing

ORGANIZATIONAL CHANGE MANAGEMENT

We like to start the implementation phase by conducting a Change Management Effectiveness Survey. The object of this survey is to determine how effective the change management activities have been during the first five phases in preparing the people for the changes that will be required. The Change Management Effectiveness Survey (CME) is designed to serve as an aid in dealing with the human aspects of an organization's adaptation to change. As a diagnostic tool, the CME can be used to determine the overall response to an organizational change and its contribution to the risk of implementation failure. The following are some typical statements:

- Do you believe that this change is really needed? Even if people fully understand the organization's rationale for a change, they may have different perspectives than that of a change sponsor and may not agree that a change is truly necessary.
- How significant do you believe the tangible, intellectual, or emotional costs of this change are for you? People resist changes that appear to be too costly relative to what they will gain.
- How compatible do you believe this change is with existing organizational values? People resist when they believe that a change introduces values that are not compatible with the existing organizational values.

- Do you believe that your daily work patterns were adequately considered when the planning was done for this change? Failing to acknowledge the impact a change may have on people's work patterns tends to promote distrust and alienation.

The inability to fully implement decisions that affect large numbers of people throughout an organization is a critical issue facing many organizations today. Major decisions with strategic implications, such as reorganizations, new policies and procedures, or a new Strategic Plan, can fail when an organization lacks the capacity to translate senior-level directives into tangible results.

One of the key factors that affects an organization's ability to implement present or future changes is its past implementation success. Previous experience is a potent predictor of the future. Prior implementation difficulties are likely to be repeated in new projects. Therefore, an assessment of previous implementation experience is critical in planning for future changes.

To do this evaluation, we suggest doing an Implementation History Assessment. The following are typical statements that would be evaluated as to how consistent the statement is with how the organization has implemented major changes in the past.

- This organization has a poor track record for identifying and resolving problems during change.
- Risk taking has been discouraged and creative ideas have been ignored because too much emphasis has been placed on finding and punishing errors.
- We usually have not allowed sufficient time for carrying out changes.
- Past change activities have been poorly monitored by managers.
- Past change efforts have not been communicated effectively down through the organization, resulting in people feeling confused about how the change affects them or what they should do differently.

The results of the analysis can be used to define the risks related to 25 key characteristics that are critical to successful implementation of change within an organization. These results will be used to help define the megatrends of the changes that can be included in the Strategic Plan. Mitigation plans should be developed for all the high risk barriers to successful implementation of the Strategic Plan and the Organization Alignment Plan.

Figure 9.2 shows typical results from an Implementation History Assessment.

Most Organizational Alignment projects require a very effective Organizational Change Management process that is applied through the Organizational Alignment Cycle. If it is not stated during Phase I of the cycle, many things can go wrong.

> We find that it is most important for a company to focus on change in the customer.
>
> **—Craig R. Stokely, founder, Stokely Partnership, Inc.**

CMT	Average	1	2	3	4	5	6	7	8	9	10
1. Layers of Approval	8.71								X		
2. Problem Solving	8.00								X		
3. Risk Taking	5.29				X						
4. Responsibility/Authority	4.43			X							
5. Middle Mgrs Involvement	5.57				X						
6. Time Allowed	8.00								X		
7. Consequences	5.43				X						
8. Discipline	5.29				X						
9. Follow-up	4.71			X							
10. Incentives	7.57							X			
11. Clarity of Communication	7.00							X			
12. Expressing Opinions	6.86						X				
13. Bureaucracy	6.71						X				
14. Teamwork	4.86			X							
15. Persistence	7.00							X			
16. Influence	6.71						X				
17. Pleasing People	8.00								X		
18. Compartmentalization	6.29						X				
19. Support	7.57							X			
20. Walking the Talk	6.57						X				
21. Low Expectation	6.57						X				
22. Vision/Strategies	7.00							X			
23. Committees	6.86						X				
24. Turf/Control	8.29								X		
25. Consensus	7.00							X			
Average	**6.65**	Low Risk			Caution			High Risk			
I.H.F	**66.51**										

FIGURE 9.2

Typical Implementation History Assessment results.

Murphy's Laws Related to Change Management

There are a number of Murphy's laws that can be applied directly to Organizational Change Management. Some of them are listed here:

- Nothing is as easy as it looks.
- Everything takes longer than you think.
- If anything can go wrong, it will.
- A day without crisis is a total loss.
- Inside every large problem is a series of small problems struggling to get out.
- The other line always moves faster.
- Whatever hits the fan will not be evenly distributed.
- Any tool dropped while repairing a car will roll underneath the exact center.
- Friends come and go, but enemies accumulate.
- The repairman will never have seen a model quite like yours before.
- The light at the end of the tunnel is the headlamp of an oncoming train.
- Beauty is only skin deep, but ugly goes clear to the bone.
- Murphy was an optimist.

THE SEVEN PHASES OF THE CHANGE MANAGEMENT METHODOLOGY

> Change management approaches may sound like common sense but, too often, common sense is not commonly practiced.
>
> **—HJH**

To offset the many problems that occur if the affected employees aren't made a part of the project before it's put into place, a seven-phase change management methodology has been developed that starts as soon as the project team has been assigned (see Figure 9.3). Following are more details related to each of the seven phases.

Phase I—Clarify the Project

In Phase I the project team defines the scope of the project and the level of commitment by management and the affected employees required for the project to succeed.

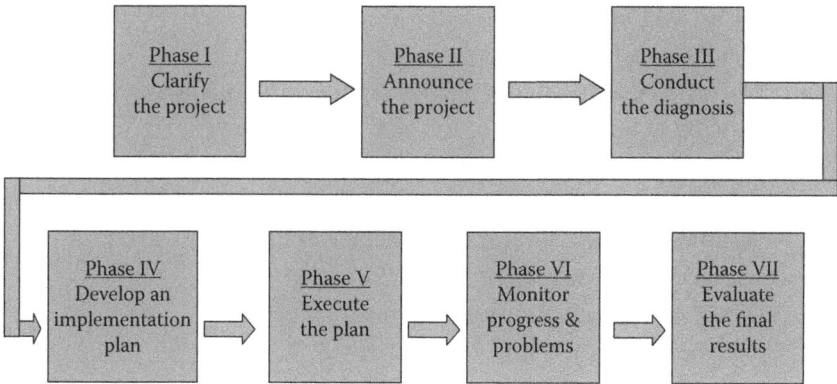

FIGURE 9.3
The seven phases of the change management methodology.

Phase II—Announce the Project

In Phase II the team develops a tailored change management plan and communicates it to all the affected constituents. Careful planning and sensitivity to the unique needs of various groups will minimize disruption and set the stage for acceptance of the need for the change.

Phase III—Conduct the Diagnosis

During Phase III the team uses surveys and other types of analysis tools (e.g., landscape survey) to determine any barriers that could jeopardize the success of the change. This diagnostic data, coupled with the rich dialogue that occurs during Phase II, provides the basis for developing an effective implementation plan.

Phase IV—Develop an Implementation Plan

The implementation plan defines the activities required to successfully carry out the project on time, within budget, and at an acceptable quality level. Typical things that will be addressed in this plan are the following:

- Implications of status quo
- Implications of desired future state
- Description of the change
- Outcome measures
- Burning platform criteria
- Comprehensive or select application of implementation architecture

- Disruption to the organization
- Barriers to implementation
- Primary sponsors, change agents, change targets, and advocates
- Tailoring of announcement for each constituency
- Approach to pain management strategies
- Actions to disconfirm status quo
- Alignment of rhetoric and consequence management structure
- Management of transition state
- Level of commitment needed from which people
- Alignment of project and culture
- Strategies to improve synergy
- Training for key people
- Tactical action steps
- Major activities
- Sequence of events

Phase V—Execute the Plan

The goal of Phase V is to fully achieve the human and technical objectives of the change project on time and within schedule. It's designed to achieve these objectives by reducing resistance and increasing commitment to the project.

Phase VI—Monitor Progress and Problems

The goal of Phase VI is to keep the project on track by consistently monitoring results against the plan.

Phase VII—Evaluate the Final Results

The intent of Phase VII is to provide a systematic and objective collection of data to determine if the tangible and intangible objectives of the project have been achieved and to provide insight into lessons learned and potential problem areas that may arise in future change projects.

CHANGE MANAGEMENT TOOLS

Change management includes almost 50 unique change management tools. Some of them are:

- Cultural assessment

- Landscape surveys
- Change agent evaluation
- Change history survey
- Change resistance scale
- Overload index
- Predicting the impact of change
- Role map application tool
- When to apply implementation architecture

Figure 9.4 indicates where these tools can be used in each of the seven phases of the change management methodology. The book *Project Change Management* (McGraw-Hill, 2000) by Daryl R. Conner, Nicholas L. Horney, and H. James Harrington provides detailed information on each of these 50 change management tools. The surveys and evaluations that make up the change management tools were developed by Daryl R. Conner and his team at ODR, which is now called Conner Partners.*

People who welcome change make progress. People that fight it make excuses.

—HJH

AREA ACTIVITY ANALYSIS (AAA)

Everyone in IBM has customers, either inside or outside the company, who use the output of his or her job; only if each person strives for and achieves defect-free work can we reach our objectives of superior quality.

—John R. Opel, past Chairman of the Board, IBM

It seems logical to us that if an organization has a mission statement and a set of performance goals, the mission statement of each of the vice presidents (VPs) of the organization has to reflect and directly support the corporate mission statement. It therefore follows that the middle management's mission statement and performance goals need to reflect and directly support the part of the related VP's mission statement and performance goals that the VP has assigned to that middle manager. Furthermore,

---

* You can buy these tools by contacting Conner Partners, 1230 Peachtree Street, Suite 1000, Atlanta, GA 30309, (404) 564-4800, www.connerpartners.com.

		Phases							
OCM Assessments, Planning Tools, and Training	**Pre-Work**	**I**	**II**	**III**	**IV**	**V**	**VI**	**VII**	
Change Agent Evaluation (**A**)	x			x				
Change Agent Selection Form (**A**)	x			x				
Change History Survey (**A**)*		x						
Change Project Description Form (**P**)	x	x	x	x	x	x	x	x
Change Resistance Scale (**A**)				x				
Communicating Change: Project Analysis (**P**)			x					
Communicating Change: Constituency Analysis (**P**)			x					
Communicating Change: Statement Development (**P**)			x					
Communicating Change: Announcement Plan (**P**)			x					
Culture Assessment (**A**)				x				
Culture Audit (**A**)				x				
Expectations for a Successful Change Project (**A**)		x						
Implementation Plan Advocacy Kit (**P**)					x			
Implementation Plan Evaluation (**A**)					x			
Implementation Problems Assessment (**A**)				x				
Landscape Survey (**A**)*		x		x		x	x	x
OCM Training for Sponsors, Agents, Targets, and Advocates (**T**)	x	x			x			
Organizational Change Implementation Plan (**P**)					x	x	x	x
Overload Index (**A**)*		x				x		
Pain Management Strategies: Sponsor (**P**)	x							
Postmortem Process **								x
Predicting the Impact of Change (**A**)		x		x				
Preliminary Implementation Plan (**P**)					x			
Role Map Application Tool (**P**)	x	x	x	x	x	x	x	x
Senior Team Value for Discipline (**A**)		x						
Sponsor Checklist (**A**)		x		x				
Sponsor Evaluation (**A**)		x		x				
Synergy Survey (**A**)		x		x				
When to Apply Implementation Architecture (**A**)		x						

Pre-Work—Used before starting Phase I.
* *This assessment tool is scored by ODR's Diagnostic Services.*
** *This project-effectiveness evaluation tool is not OCM specific.*

FIGURE 9.4

OCM assessments, planning tools, and training.

the same condition applies to the interface between middle management and the first-line managers, supervisors, or natural work team leaders. Under no circumstances can a higher level manager delegate to a manager who reports to him or her an authority that he or she doesn't have. This rule is called the Cascading Mission, Authority, Responsibilities and Measurement rule, which all organizational structures are based upon. Area Activity Analysis (AAA) is the best way we have found to implement this rule throughout the organization.

Area Activity Analysis is a proven approach used by each natural work team (area) to establish efficiency and effectiveness measurement systems, performance standards, improvement goals, and feedback systems that are aligned with the organization's objectives and understood by the employees involved. We believe that AAA is the most effective and efficient team-related approach developed to date, as it relates the individual's assignment back to the organization's objectives. It also establishes the total measurement system down to the individual department level.

As viewed by the employees, the organization chart often looks like everyone above them exists just to generate work for them, and all that management wants from them is their complete discipline.

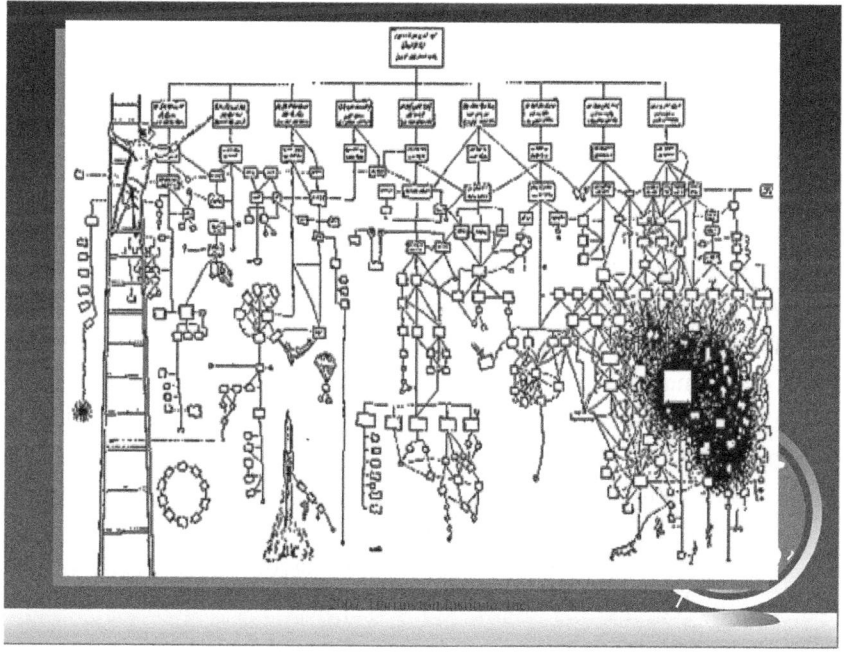

Since the early 1960s, management has unsuccessfully chased after the elusive pot of gold at the end of the rainbow called "Internal Supplier-Customer Relationships." This concept is a sound and relatively simple one, but it is extremely difficult to implement and maintain. Theoretically, treating the person who receives your output the same as you would treat an external customer is relatively straightforward. Unfortunately, people are less inclined to react to a person who is not paying for their output than to individuals who show the value they place on their output by paying directly for it.

We hear a great deal these days about the importance of developing the internal customer-supplier relationship within our organizations. No matter where you go, people profess to be concerned about satisfying their internal customers. Every quality training program teaches it, every book on quality preaches it, and clearly most people today truly believe internal customer satisfaction is critical to business success. The truth of the matter is that it is very difficult for any organization to have good external customer satisfaction unless it has excellent internal customer satisfaction.

> Definition: *Natural Work Team (NWT)* or *Natural Work Group (NWG)* is a group of people who are assigned to work together and report to the same manager or supervisor.

AAA will help you:

- Clarify your NWT's real purpose.
- Identify those time-consuming activities that do and do not support your mission.
- Bring better alignment between your mission, activities, and the expectations of your internal and external customers.
- Align your employees' activities with the NWT's priorities.
- Identify which activities add real value and which can be minimized or eliminated.
- Understand how to make the transition from finding and fixing problems to preventing them.
- Clarify your requirements for your internal and external suppliers and measure their performance.
- Define a comprehensive measurement system for the critical activities that take place within your NWT and set performance standards for each of them.
- Put together an implementation plan to make it all happen.

Many of us are frustrated with our inability to accurately put a finger on the source of the problems we face every day. We know we work hard, deliver quality output, and drive the people who work for us to do the same. The harder we work and the more we push, the more entrenched the problem seems to get and employees' morale seems to sink. It is indeed a vicious, negative cycle that can sap the energies of even the most dedicated manager or employee. Here again AAA can help.

- AAA is a simple but powerful tool that you can begin using immediately to provide clear direction on what may otherwise become a confusing journey toward improving customer service. AAA aligns every unit with the organization's Strategic Plan.
- AAA can help you align your energy and resources with your organization's mission in a way that can result in greater effectiveness, efficiency, satisfaction, and teamwork.
- AAA serves as a compass to help you find your way through the jungle of overwork that threatens to overtake us all.
- AAA helps you sort out the vital few from the trivial many so that you can focus on delivering the value that you, and you alone, can add to your organization and your customers.

AAA was designed by managers as a way to analyze and organize their work areas to get better results from their current resources. As the weeks, months, and years go by, the organizations we manage inevitably take on more and more responsibilities and our jobs get more and more complex. We begin to feel like we're running on a never-ending treadmill while someone keeps turning up the speed.

AAA is unlike any other technique for improving processes, reducing costs, or decreasing turnover. It helps everyone clarify expectations and focus their efforts on the area's mission.

AAA is an appropriate tool for new or existing areas or departments. It is a tool that will help to ensure that everyone understands their area's mission, customer expectations, what they need to do to succeed, and how to measure their performance.

AAA can be used by managers at any level of the organization to improve the efficiency, effectiveness, and teamwork within their operations. It can be used by an individual unit or as part of a coordinated, organization-wide effort. It can also be used by an individual to improve his or her performance. It can and should be used by every person in the total organization at every level, from the team of employees (vice presidents) who report directly to the president of the organization to the team of maintenance workers who report directly to the maintenance line manager.

What Is Area Activity Analysis (AAA)?

AAA is the first performance improvement tool that a manager should use to help his or her area get started on a sound footing. AAA helps the organization accomplish the most basic of all management tasks, which is to define the following:

- The purpose of each area
- What activities must be done to satisfy the area's mission
- The area's internal and external customers' requirements
- How the area's performance should be measured
- What acceptable performance is

AAA is not another technique for improving processes, lowering cycle time, or reducing costs. Simply stated, it helps managers clarify what is expected of their groups, by defining the department's key measurements, called Department Performance Indicators (DPIs); setting performance standards; and focusing people's efforts like a laser beam on the organization's mission. It defines whether the area needs to improve and where the improvement opportunities exist. It helps the employees understand what is important for their customers and managers. It also helps the employees understand how they fit into the organization and contribute to the organization's goals. It is a people-building approach that helps them stand on their own feet with a high degree of confidence in themselves and helps them understand that they are doing something worthwhile.

Before we go any further, let's discuss what we mean by customer-supplier relationships as they relate to internal and external customers. Basically, a customer-supplier relationship can develop in two different ways.

1. An individual or organization can determine that it needs something that it does not want to create itself. As a result, it looks for some other source (supplier) that will supply the item or service at a quality level, cost, and delivery schedule that represents value to the individual or organization (customer).
2. An individual or organization (supplier) develops an output that it believes will be of value to others. Then the individual or organization looks for customers that will consider the supplier's output as being valuable to them.

Two key points need to be made:

- A customer-supplier relationship cannot exist unless the requirements of both parties are understood and agreed to. Too often, customers expect input from suppliers without understanding their requirements and/or capabilities and without defining exactly what they need. On the other hand, too many suppliers provide output without defining their potential customer's requirements and obtaining a common, agreed-to understanding of what both parties' requirements are.

- Both the customer and the supplier have obligations to provide input to each other. The supplier is obligated to provide the item or service and define future performance improvements. The customer has an obligation to provide compensation to the supplier for its outputs and feedback on how well the outputs perform in the customer's environment.

The customer-supplier process has a domino effect. Usually, when a supplier is defined, that supplier requires input from other sources in order to generate the output for its customer. As a result, it becomes both a customer and a supplier (see Figure 9.5).

Although the procedures related to internal customer-supplier partnerships are less stringent and have been simplified because the internal customer does not pay for the services that are provided, the concepts are equally valid. Too often, we set different standards for internal suppliers than we have for external suppliers. As a result, many of the internal suppliers provide outputs that are far less valuable than it costs to produce the outputs. This often results in runaway costs and added bureaucracy. With AAA, we will show you how to apply the customer-supplier model to the internal organization, thereby improving quality and reducing cost and cycle time of the services and items delivered both within and outside the organization. It also serves to bring each area within the organization in line with the Strategic Plan.

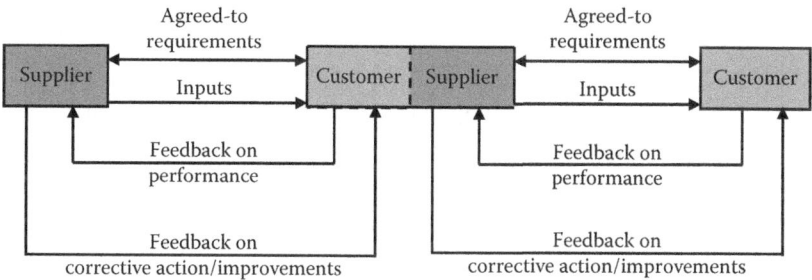

FIGURE 9.5
Feedback on corrective action/improvement.

Phase	Number of Activities
Phase I Preparation for AAA	5
Phase II Develop Area Mission Statement	6
Phase III Define Area Activities	8
Phase IV Develop Customer Relationships	7
Phase V Analyze the Activity's Efficiency	6
Phase VI Develop Supplier Partnerships	5
Phase VII Performance Improvement	8
Total	**45**

FIGURE 9.6
The seven phases that make up the AAA methodology.

Defining customer-supplier relationships is only one part of making an area function effectively. There are many other factors that also must be considered. For example:

- What is the area responsible for?
- How is the area measured?
- What is acceptable performance?
- How does the area fit into the total organization?
- How well do the area's employees understand their roles and the ways they can contribute?
- What are the important activities that the area performs from top management's standpoint?
- How does the area align itself with the organization's Strategic Plan?

It is important to note that all of these questions pivot around the activities that the area is involved in. It is for this reason that the AAA methodology has a broader perspective that goes beyond the customer-supplier partnership concept to embrace a complete business view of the area.

The AAA methodology has been divided into seven phases (see Figure 9.6) to make it simple for the NWT to implement the concept. Each of these phases contains a set of activities that will progressively lead the NWT through the methodology.*

---

* For more information on Area Activity Analysis see H. James Harrington, *Area Activity Analysis* (New York: McGraw-Hill, 1998).

ORGANIZATIONAL ALIGNMENT AND COMMUNICATIONS

Communication of purpose, strategy, processes, structure, culture and people alignment is as important as creation if employees at all levels in all functions, as well as external people associated with the organization, are going to understand its direction and management practices. Unless the purpose is communicated effectively, it cannot become a driving force in corporate activity.

Effective communication and alignment of purpose, strategy, processes, structure, culture, and people depends on a few simple principles:

1. A clear purpose that captures the attention of the audience takes careful planning and the dedication of a person who champions the statement.
2. Multiple and periodic communication of the purpose of the Organizational Alignment strategy improves the chances of its being understood.
3. Evaluation of the results of the communication effort provides a basis for improving and reinforcing it.
4. Beyond the dissemination of the alignment strategy, management must also support its purpose in all its actions and words and ensure that all procedures and structures within the organization support the alignment. Otherwise, communication will simply be pushing what people perceive as meaningless propaganda.

Managers should ask, "What do we want our communication of the alignment purpose statement to accomplish?" Their objectives should be as specific and concrete as possible. One approach is to establish short-, medium-, and long-range objectives. Then managers can evaluate the effectiveness of the process in meeting these objectives. Short-range objectives cover initial communication of the purpose statement. Here are some examples of short-range objectives:

1. Share the purpose of the Organizational Alignment with all employees and appropriate external people.
2. Generate excitement and interest in the organization's purpose.

3. Solicit ideas for fulfilling the purpose of the new alignment.
4. Communicate a direction and challenge in ways that meet individuals' various needs.

Definition: *Communication Diagram* is a pictorial view of the way information is transferred within an organization.

The communication diagram is a very useful tool that defines the way people communicate with each other as defined by the organizational structure. It also helps to define levels of authority and responsibility, as well as the formal interface between the different operational units. Figure 9.7 is a communication diagram for a knowledge management system.

Effective transitions are delivered through an alignment of business objectives, organizational communications, and employee engagement. The Organizational Alignment methodology takes leaders and employees

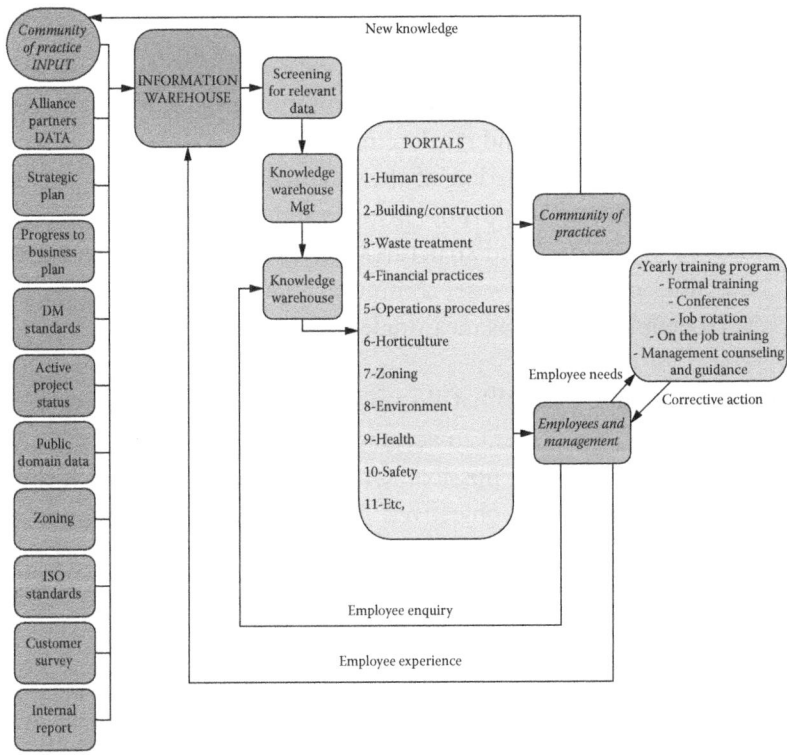

FIGURE 9.7
Technical communication diagram.

through an interactive information gathering process and then offers insights on how to create new strategic alignments that foster increased productivity and performance. "Leaders need an efficient way to gather feedback, assess communications and develop strategies that address identified gaps in organizational performance and alignment," explained Doug Griffen, director of strategy and facilitation and founder of the Advanced Strategy Center. DiscoveryBridge* is a dynamic methodology to take the pulse of an organization and quickly get focused on what needs to be done to create sustainable productivity and achieve Organizational Alignment, in order to create a communication diagram.

Key Implementation Issues

After establishing objectives for communication, management has to identify who should receive what information, which is a major implementation issue. The list should be both comprehensive and at the same time Lean. To do this successfully, we recommend dividing the audience into two broad categories—internal and external—and then segmenting each category further, according to the information it needs and will use and its members' abilities to comprehend abstract and specific goals. Communication in one division, for example, may emphasize a narrower body of information than what corporate people are told. A communication about the statement of purpose for supervisors may supply less detail than one for top management.

Vary communication of the purpose statement to the degree that distinct needs and views among subgroups may affect how they understand or apply the purpose statement. One executive recently told us that subgrouping forced his organization to consider what each group would think about and how they would react. In short, it helped them be more purposeful in their communication.

---

* The following is a synopsis of the three-phase DiscoveryBridge methodology: (1) Discovery I addresses the perspectives of leadership and senior function. Discovery II is a parallel process that aligns the input of the employee audience (or other stakeholder group) to the input of the leadership team. (2) The second DiscoveryBridge phase, Engagement, is a strategic analysis of the organizational goals and an assessment of the barriers measured against DiscoveryBridge's leadership principles: (a) to foster a culture of inclusion, (b) nurture organizational pride, (c) exhibit social consciousness, (d) define a leadership legacy, and (e) support constant discovery. Out of this phase comes an "organizational atlas." (3) The third phase, Performance, helps implement an overall implementation strategy that utilizes Porter Novelli's stakeholder management approach, all with the goal to measurably improve levels of sustainable performance. Additional information on the DiscoveryBridge technology and methodology can be found at http://www.discoverybridge.com.

Gauging Effectiveness

To create the future, and to effectively implement communication and feedback channels within an organization, the organization must be capable of engineering the entire end-to-end value stream. Experienced managers know to redefine business designs and processes when implementing new forms of value. What's different in this new environment is the widespread use of synergistic communications and feedback clusters, business ecosystems, coalitions, cooperative networks, or sourcing to create end-to-end value streams. These feedback systems link businesses, customers, and suppliers to create a unique business organism.

> The business design of the future increasingly uses reconfigurable communication models to best meet customers' needs.
>
> **—James Carlopio, PhD, Australian Graduate School of Management,**
> **University of New South Wales**

In gauging the effectiveness of previously cited success stories of e-commerce startups such as Amazon, Microsoft CarPoint, and E*TRADE, there are many implementation issues involving essentially complex structures around which the ultimate effectiveness can be gauged. The effectiveness of these complex structures are gauged as to how well they meet their purpose of organizing and energizing cross-enterprise relationships to create end-to-end value for the customer. In other words, the customer is the arbiter of success. Communication strategists see organizations like these as being the center and alignment hub of an extended business family that pools the resources and benefits of each organization's expertise. An effective feedback and communications system can be gauged by how well it plays a role in attacking market leaders, and new entrants (and established organizations as well) are using communications tools such as the previously mentioned DiscoveryBridge to gain access to resources, customers, technology, and products.*

---

* DiscoveryBridge identifies obstacles leaders face in addressing alignment of employee engagement, communications, and performance. Following 10 years of research that identified barriers to employee engagement, breakdowns in organizational communications, and a new defining set of leadership principles, the Advanced Strategy Center and PNConsulting unveiled a proprietary employee engagement methodology and technology, DiscoveryBridge, to more than 100 Arizona business leaders who were briefed in Scottsdale on emerging new management methodologies.

In his paper "Strategy Implementation—An Insurmountable Obstacle?" Andreas Raps itemized the strategy implementation's top 10 checklist as follows:

- Commitment of top management
- Involve middle manager's valuable knowledge
- Communication is what implementation is all about
- Integrative point of view
- Clear assignment of responsibilities
- Preventive measures against change barriers
- Emphasize teamwork activities
- Respect the individuals' different characters
- Take advantage of supportive implementation instruments
- Calculate buffer time for unexpected incidents

ALIGNMENT TOOL #7: COMMUNICATIONS SURVEY

We find the following survey a very useful way to evaluate the communication activities.

Instructions: Please read each statement and indicate the extent to which it describes the reality in your organization as a whole. Your responses should reflect what you have experienced as well as what you have generally observed.

Answer each question using a 10-point scale. The left side of the scale indicates that you *totally disagree* and the right side that you *strongly agree*. If you do not know, indicate so next to the question.

Please take the time to respond to the open-ended questions at the survey end, as your responses are critical in improving the readiness for change. Be honest in your responses as there are no right or wrong answers. Your responses will remain completely confidential.

1. My organization has effectively communicated mission, vision, and values statement(s).
2. Managers anticipate coordination requirements and interdependencies when planning work activities.
3. Our organization has effective lateral coordination and communications the way it works.
4. People at all levels communicate and cooperate with one another.
5. The organization encourages networking and open, multidirectional communications rather than a traditional vertical channel approach.

6. Every employee understands what is expected in terms of performance and receives regular feedback on how he or she is doing.
7. The communication procedures and channels in our organization are effective.
8. All employees engage in open and honest communications.
9. Managers discourage guarded and cautious communications at all levels.
10. Internal communications and day-to-day activity support the organization's purpose and plans.

Totals: Add up the scores for each of the 10 questions: _____

Note: Grand Total of 50 or less indicates very serious problems and 80 to 100 indicates no serious issues, while 50 to 80 indicates need for improvement.

There are a number of innovative new ways in which you can communicate with your employees. Some questions pertaining to these methods are the following:

- How well does your organization perform instant messaging? Almost every Web browser has an instant messaging feature, and this can be a faster method of communicating than even e-mail. Because the instant message alert pops up on the screen each time there is a message, this can be a quick way to communicate with employees working in the office as well as remotely.

  _____
  _____
  _____
  _____

- How well do you control the option of your employees using instant messaging to chat about personal matters when they should be working?

  _____
  _____
  _____
  _____

- How well do you use Snap Comm tools? One of the more unique methods of employer-employee communication on the market today, Snap Comm tools feature six distinct features that allow employers to get their message out to their employees without having to send a slew of individual messages.

  _____

  _____

  _____

  _____

- Do you use Snap Polls or Snap Quiz? Want to know exactly what your employees think on a variety of issues? Let them take this quick and easy on-screen quiz that boasts an improved response rate because of its ease of use. On-screen quiz features will help you get an idea of where your employees might be weak in terms of organization policy. There is an automated scoring option that takes the work out of getting all of the results.

  _____

  _____

  _____

  _____

- Does your organization use Snap Ticker? Similar to what you see on CNN, these scrolling tickers will give your employees critical updates and disseminate important organization information with the click of a button.

  _____

  _____

  _____

  _____

- Do you use Snap Shots? These screensavers act as billboards, broadcasting your message and ensuring that it is being reinforced every time your employee is forced to view it. A great way to communicate some core organization concepts.

  _____

  _____

  _____

  _____

THE MANAGEMENT CYCLE AND ALIGNMENT

As can be seen in the studies by alignment experts comparing the scores of the "best" and "worst" organizations, the management cycle score clearly drives the final effect on the management performance score. The visible difference between "best" and "worst" organizations is clear: proactive versus passive; perceptive versus reactionary; energizing versus draining; and feed-forward versus feedback.

As outlined by Sericho Yahagi, there are 12 factors and some 41 subfactors involved in the whole system integration aspects of Organizational Alignment.* These factors, along with their associated subfactors, are as follows:

- Corporate History: past, present, and future
- Corporate Climate: core, climate, and culture
- Strategic Alliances: objectives and coherence
- Channels: suppliers and buyers
- Management Cycle: vision, strategy, planning, organizing, implementing, and controlling
- Environment: economic, societal, and global
- Total Improvement Management (TIM) targets: inputs, markets, technologies, and products
- Business Structure: business fields, business mix, and market standing
- Management Resources: money, materials, information, and people
- Management Design: system, organization, authority, and responsibility
- Management Functions: decision making, interrelationships, and quality
- Measurement System Performance: growth, scale, stability, profit, and market share

Of the 12 factors, 6 are critical to the success of a company's Organizational Alignment: the management cycle, business structure, management resources, management design, corporate culture, and measurement system performance. These six factors are interrelated in a

---

* According to Yahagi, the circulatory aspects of the management quality system are a key to understanding the relationships of this system to others, such as the Malcolm Baldrige National Quality Award.

circulatory system of factors which have a cause-and-effect relationship as follows: The management cycle acts as the driver and influencer of the four factors of structure, resources, design and culture—which in turn affect measurement system performance.*

If the quality level of the management cycle is low, then the six factors generate a negative or bad feedback loop, which finds management passively waiting until poor results of management performance forces reactionary feedback into the management flow. If the quality level of the management cycle is "high," then these six factors generate an excellent feed-forward loop in which management perceives the strategies and plans needed for success of each management factor and then proactively formulates and implements them.

In other words, a "bad" or poor management cycle drains the quality from the other management factors, and an excellent management cycle pumps up the quality of the other four factors. Thus, the synthesized effects of these four factors cause the "score" of the measurement system performance to be either higher or lower than the management cycle, depending upon the good or bad conditions present. Finally, because management performance flows into the management cycle, the cyclical corporate growth of the "best score" organizations will incrementally improve the management cycle and, ultimately, overall management quality.

Of particular interest and importance are the responses for the first two elements in the management cycle: corporate vision and management strategy. In both cases, the highest number of respondents scored the maximum of 5 points for both categories. In the case of the vision, there is a distinctive organizational corporate vision, which is stipulated as a long-term management direction. In the area of strategy, a comprehensive management strategy exists which has been created from the vision and has been linked to planning. As previously outlined, a compelling vision or outcome that is well articulated, shared by senior leaders, and effectively communicated down through the organization is truly the most valuable asset on which all of the successful alignment depends and with which it begins. A company vision must be "believable" in that employees perceive it as feasible and worthy of people's time and effort, without seeming overwhelming.

---

* For further details, see Seiichiro Yahagi, "After Product Quality in Japan: Management Quality," *National Productivity Review* (Autumn 1992): 501–515.

Organizations with a well-understood vision that is internalized by staff employees, as well as managerial employees, have senior leaders who recognize that from their vantage point they cannot always see where the communication gaps are in the organization, so they manage this explicitly. In organizations where senior leadership is not aligned around the same vision, this is felt throughout the organization, and it may find its key groups working at different purposes.

Depending on the organization and the situation, an effective implementation plan may lead to and involve all of organizational development and training activities outlined below, integrated into a cohesive Organizational Alignment plan. Each of these areas will produce significant outcomes, if enough attention is given to the area. The question becomes "Which areas in the management cycle should we focus on first?" and then the desired outcomes can be prioritized.

- Organizational Alignment Plans
- Culture Building Initiatives
- Customer Service Consulting Training Plans
- Internal Communication and Internal Branding Plans
- Workshop and Retreat Design and Facilitation
- Rollout Strategy
- Intranet and Internet Database Development
- Creation of Field Training Tools and Activities
- Training Module Instructional Design
- Training Delivery
- Innovation Workshop Facilitation
- Corporate "University" Creation
- Recruiting and Hiring Strategies
- Orientation Program Development
- e-Learning/Distance Learning Applications

Key Implementation Challenges

Like some effective redesign efforts, involving the whole system management cycle produces an easily understood payoff by almost immediately producing positive changes to the organization's bottom line. This is most often a result of corporate realignment and downsizing via restructuring, as well as the impact upon the supply chain of the organization. In addition, redesign using the whole system model has the positive effect of changing a firm's work orientation from bureaucratic, political and disjointed to one that is more synchronized, uniform, and efficient.*

Some experts claim Business Process Improvement (BPI) has too much "top-down" orientation of planning and execution and, as a result, it fails to fully consider all elements of the organizational system (such as customers or suppliers). The Organizational Alignment approach integrates BPI with a total integrated management cycle whole systems approach obtaining the breakthrough results. A number of organizational theorists feel the BPI approach fails to adequately emphasize *where* the organization wishes to go with its transformation alignment efforts and *who* is involved in getting it there. The management cycle whole system approach places primacy upon addressing these issues and was initially developed by MaxComm Associates with Strategy Associates to facilitate organizational transformations and alignment in client organizations.

ORGANIZATIONAL ALIGNMENT AND BUSINESS PERFORMANCE MANAGEMENT

Business Performance Management (BPM) is a key business initiative that enables organizations to align strategic and operational objectives with business activities in order to fully manage performance through better informed decision making and action. BPM is generating a high level of interest and activity in the service business and IT community because it provides management with their best opportunity to meet their business measurements and achieve their organizational goals. IBM uses the term "BPM" for business initiatives that emphasize aligning strategic and operational objectives in addition to monitoring and managing business processes and associated IT events. In this section, we describe BPM in detail, review BPM trends, and introduce a BPM framework and implementation methodology.

BPM (sometimes called Corporate Performance Management, Enterprise Performance Management, or Operational Performance Management) is a set of processes that help organizations optimize their business performance. It is a framework for organizing, automating, and analyzing

---

* John Childress and Larry Senn, *In the Eye of the Storm: Reengineering Corporate Culture*, Leadership Press, NY, 1995.

business methodologies, metrics, processes, and systems that drive business performance.

BPM is seen as the next generation of Business Intelligence (BI) and helps businesses make efficient use of their financial, human material, and other resources. For years, owners have sought to align strategy down and across their organizations, they have struggled to transform strategies into actionable metrics, and they have grappled with meaningful analysis to expose the cause-and-effect relationships that, if understood, could give profitable insight to their operational decision makers.

BPM is all about taking a holistic approach for managing business performance and achieving the business goals. Businesses align strategic and operational objectives, and business activities, to fully manage performance through more informed and proactive decision making. This holistic approach enables the integration and use of business intelligence, process management, business service management, activity monitoring, Business Process Improvement, and corporate performance management to achieve a single view of their organization. Businesses are evolving to an environment capable of supporting continuous data flow, which enables the support of business intelligence environments with more current data.

One of the primary goals of BPM is getting people within the organization to do what they're supposed to do. A number of "soft" techniques can help achieve this goal: strong executive sponsorship, communication, training, change management. But in the end, the key to success is aligning all aspects of performance management with things people can understand and personally control. Alignment is a simple concept, but making it work is the most challenging and enigmatic aspect of performance management.

Many high-level performance measures are so abstract they don't mean anything to the people who actually do the work. There's nothing wrong with having a few key performance measures—in fact, they're essential—but those high-level measures need to be broken down into a set of more focused measures that are meaningful to employees at every level. For example, net cash flow might be a critical performance measure for the CEO and the organization overall—but what does it mean to an accounts receivable clerk, and what can that person do to improve net cash flow performance?

There are various methodologies for implementing BPM. It gives organizations a top-down framework by which to align planning and execution, strategy and tactics, and business unit and organization objectives. Some of these are Six Sigma, Balanced Scorecard, Activity-Based Costing, Total

Quality Management, Economic Value-Add, Area Activity Analysis, and Integrated Strategic Measurement. The Balanced Scorecard is the most widely adopted performance management methodology. Methodologies on their own cannot deliver a full solution to an organization's Critical Path Method (CPM) needs. Many pure methodology implementations fail to deliver the anticipated benefits because they are not integrated with the fundamental CPM processes.

The following is an excerpt from the article "Redefining BPM: Why Results and Performance Must Be Separated," by Harry Greene (published by Business Performance Management): "Many problems that plagued 20th-century businesses and continue to dog companies today are by-products of those businesses' lumping of the capital they utilize and the results they produce together as 'performance' in their BPM. The popular 20th-century definition of 'business' is the activity of providing goods and services. This definition includes two components: the activities of the business, which is to say its utilization of capital in actions executed, and the goods and services provided by the business as output results accomplished. In judging an organization's performance, 20th-century management mixes together these two components of the business." We propose a 21st-century definition of "business," in which the distinction is impossible to ignore:

> Definition: *Business* is an organization that utilizes capital to produce a product, provide services, or both.

Outcomes

Organizational Alignment addresses the issue by translating each high-level target into a cascading series of focused performance measures, each designed to drive specific behavior at a particular level in the organization. Using our previous example, the CEO might focus on net cash flow while the CFO looks at debt-to-equity ratio. The controller might focus on liquidity ratio, while the accounts receivable manager looks at days sales outstanding, and the accounts receivable clerk worries about percent of collections over 30/60/90 days. With CPM, employees at every level are measured by something they understand and control, and that same measure is clearly linked to the goals of their direct supervisor and the organization as a whole.

Organizational Alignment also provides shared performance measures that help align people across organizational boundaries. For instance, a performance measure that includes percent of collections over 30/60/90 days might be applied both to accounts receivables clerks and sales representatives. Shared and integrated performance measures encourage people to collaborate, boosting the organization's overall performance.

Gauging Effectiveness

Most of the time, BPM simply uses several financial and non-financial metrics or key performance indicators to assess the present state of the business versus a desired future state, and to prescribe a course of action. Some of the areas from which top management analysis could gain knowledge by using BPM include the following:

1. Customer-related numbers
   a. New customers acquired
   b. Status of existing customers
   c. Attrition of customers (including breakup by reason for attrition)
2. Turnover generated by segments of the customers—these could be demographic filters
3. Outstanding balances held by segments of customers and terms of payment—these could be demographic filters
4. Collection of bad debts within customer relationships
5. Demographic analysis of individuals (potential customers) applying to become customers, and the levels of approval, rejections and pending numbers
6. Delinquency analysis of customers behind on payments
7. Profitability of customers by demographic segments and segmentation of customers by profitability
8. Campaign management
9. Real-time dashboard on key operational metrics
10. Overall equipment effectiveness
11. Click stream analysis on a Web site
12. Key product portfolio trackers
13. Marketing channel analysis
14. Sales data analysis by product segments
15. Call-center metrics

This is more an inclusive than exclusive list. The previous list more or less describes what a bank would do, but it could also refer to a telephone company or similar service sector company. What is important are the following issues:

1. Data related to Key Performance Indicators (KPIs), Department Performance Indicators (DPIs), and Individual Performance Indicators (IPIs) that are consistent, correct, and provide insight into operational aspects of an organization
2. Timely availability of data related to KPIs, DPIs, and IPIs
3. KPIs, DPIs, and IPIs that are designed to directly reflect the efficiency and effectiveness of a business
4. Information presented in a format that aids decision making for top management and decision makers
5. Ability to discern patterns or trends from organized information

Integrating the components of performance management isn't easy. It requires collaboration, patience, and commitment across the entire organization. When Critical Path Method (CPM) is first introduced, operating units and divisions often resist, viewing integration as a threat to their decision-making independence. But the vast majority of units and divisions eventually discover that CPM is an enabling process that helps improve their decision making, laying the groundwork for the organization's future success.

Today's leading CFOs and finance organizations are implementing Integrated Performance Management to improve information quality and visibility and to generate new business insights. They are also using CPM as a tool to meet the market's increasing demands for transparency, reliability, timeliness, and accountability. With investor confidence at an all-time low and competition at an all-time high, the need for Integrated Performance Management is more critical than ever.

We believe that Organizational Alignment and change management need to be organization-wide endeavors. Presenting a "flavor of the month" training initiative to frontline associates only and expecting behavioral change is simply unrealistic; our approach, even with stand-alone customer service training programs, is to ensure that the entire organization is aligned around the training content and is prepared to practice what is preached and that the content reinforces any internal branding. It is also important to us that our clients "own" their training, so that the reinforcement of the program principles are an everyday, ongoing occurrence, not a once-a-year phenomenon.

With this in mind, organizational development, customer service consulting and training initiatives should target the following areas:

- Leadership Development
- Management Leadership, Coaching, and Feedback Training
- Job Skill Mastery
- Frontline Customer Service Excellence
- "Train the Trainer" Knowledge Transfer Modules
- Service Excellence Champion/Advocate Training
- Organizational Alignment from the Executive Suite to the Frontline
- Relationship Building in Sales and Service

OUTSOURCING

To effectively deal with the dynamic changes that many organizations are facing, the trend toward outsourcing has gained a new momentum, in that most organizations know that they cannot do everything well. There is one indispensable caveat: never outsource a core competency.* First-generation outsourcing paid attention to improving efficiency and reducing costs, many times ignoring the customer satisfaction consequences. At the time, the increasing complexity of computers and networks stimulated many organizations to consider outsourcing their technology management, with the biggest beneficiaries being the computer service organizations, such as IBM and EDS. For example, BellSouth outsourced its entire IT function to the EDS/Andersen Consulting venture worth more than $4 billion, thus moving toward service-oriented architecture (SOA).†

---

* According to our model approach, clients refrain from outsourcing their business processes in isolated blocks to different organizations. Accordingly, the higher demand will be for providers that can deliver all aspects of outsourced processes in one package.

† Service-oriented architecture (SOA) enables the customers to take one piece (or several pieces) from the process they wish to outsource and render to a provider (just like in Lego where one cube can be taken away from the whole structure and replaced by another one). Service-oriented architecture helps organizations concentrate on one piece of a process, thus enabling them to provide more detailed specialization for outsourcing vendors.

With the potential for so many problems, why are US companies still turning to offshore outsourcing as a solution? Many say it is because offshore developed software is not only cheaper, but it has higher quality.

—David Vogel and Jill Connelly, *Handbook of Business Strategy,* **2005**

However, the end of 2008 caused aggravation and tension in the competition between outsourcing organizations. There are lots of offshore vendors nowadays, and new organizations emerge every day. And in the situation that has formed at the present moment, a considerable number of outsourcing organizations have been unable to survive. That's why experts predict some significant consolidations like EDS/Andersen in the near future, and the outsourcing boom extends well beyond computers. Starting in the 4th quarter of 2008 and extending well into 2009, outsourcing in the form of contract manufacturing has caught on as well, as organizations searched for new and innovative ways to cut costs. Examples in the high-tech industry include Solectron, Flextronics, and SCI Systems. Outsourcing is changing the nature of the relationship between contract manufacturers and the original equipment manufacturers (OEMs).

Globalization of the sector is caused by the outsourcing boom, and historically the main outsourcing countries have been India and China. They still remain so, but a number of other countries have joined the outsourcing movement, for example, countries of Eastern Europe (Russia, Ukraine). So while the outsourcing sector gains the global scale, global organizations have started taking note of their impact on the environment. Thus it can make them change their demands on their suppliers. This is the reason why lots of outsourcing engagements now include a paragraph on the environmental protection. Environmental concern is becoming one of the evaluation criteria for service providers, and environmentalism in terms of green is becoming a trend.* The large number of outsourcing organizations leads to the more definite specialization of outsourcing. If the objective is to please customers, the best relationship for both parties

---

* Brown-Wilson Group's 2008 Outsourcing Study, June, 2008, Tampa, FL. The annual State of the Outsourcing Industry report from the Brown-Wilson Group showcases the results of 5 years of outsourcing research and up-to-date insight from 24,000 outsourcing customer survey respondents from 40 countries. The 60-page report includes the top performing vendors, advisors and consultants, locations, and trends, and looks forward at what's next for the evolving, dynamic, and often controversial industry. Appearing in the report again in 2008 was a compilation of the top outsourcing vendors, as well as outsourcing advisor and consultant organizations that excelled at providing value and process transformation to the global business community

is to behave as a single company—truly cooperative and aligned around some basic policies. This means that organizations have to share sensitive design information, link internal applications systems, and provide shared services throughout the policy-deployed supply chain.*

In Phase III, Organizational Structure, one key improvement approach was the outsourcing of non-core processes. This requires the organization to establish a new kind of working relationship with many of its suppliers.

> Only 5% of the organizations in the West truly excel. Their secret is not what they do—it is how they do it.
>
> —HJH

Key Implementation Challenges

In a growing number of cases, outsourcers finish the product, put on the logo, and ship it to the user or distributor, which creates many implementation challenges. A well-aligned Policy Deployment program helps them to achieve these objectives. Policy Deployment is considered by some Organizational Alignment experts as the wave of the future. As organizations face complex business challenges, they increasingly farm out many tasks to cut down on time to market. Increasingly, new entrants use outsourcing alliances as a business model to gain market position against a leader. The implementation challenge is often called GBF, "get big fast." This new generation of outsourcing alliances is called a variety of names, including aligned communities, clusters, and coalitions. While successful deployments differ widely from industry to industry, a common thread runs through them. They all seek to nullify the advantages of the leader by using deployment methods to achieve speedy implementation and quickly create reputation, economies of scale, cumulative learning, and preferred access to suppliers or channels.

ALIGNMENT TOOL #6: READINESS SURVEY

The following is a readiness survey format that we have found useful.

---

* While the cultures of alliance-based organizations can be very stable over time, they can never be accused of being too static. A crisis will sometimes force a group of organizations working together to reevaluate some values or set of practices. New challenges often lead to the creation of new ways of doing things. Turnover of key members, rapid assimilation of new employees, diversification into very different businesses, and geographical expansion can all weaken or change the culture throughout the value chain.

Instructions: Please read each statement and indicate the extent to which it describes the reality in your organization as a whole. Your responses should reflect what you have experienced as well as what you have generally observed.

Answer each question using a 10-point scale. The left side of the scale indicates that you *totally disagree* and the right side that you *strongly agree*. If you do not know, indicate so next to the question.

Please take the time to respond to the open-ended questions at the survey end, as your responses are critical to improving the readiness for change. Be honest in your responses as there are no right or wrong answers. Your responses will remain completely confidential.

1. Organizational leadership has identified the key stakeholders and used their requirements to set the direction of the organization.
2. Leaders role-model the purpose, vision, and values by consistently demonstrating and communicating them, thus reinforcing the direction.
3. Senior leaders are personally and visibly involved in creating an environment where legal and ethical behavior is routinely practiced.
4. Empowerment, innovation, agility, and learning are part of the organizational planning process.
5. Corporate governance processes are clearly defined and routinely followed.
6. The organizational Performance Review process is aligned and integrated with the Strategic Planning process.
7. The organization engages the leaders in a systematic planning process to develop and deploy its strategy, which includes stakeholder needs, customer requirements, external environment and competition, and internal capabilities.
8. There are specific short-term and longer-term planning horizons in place and clear criteria for the horizon chosen.
9. There are specific objectives and goals, which are aligned and balanced among financial performance, human resource development, process improvement, and customer results.
10. Strategy is deployed at every level of the organization through goals and objectives, which link and align from the organizational level all the way down to each and every individual contributor.

Totals for Reality Check: Add up the scores for each of the 10
   questions: _____
Note: Grand Total of 50 or less indicates very serious problems and
   80 to 100 indicates no serious issues, while 50 to 80 indicates
   need for improvement.

Open-Ended Questions

- Where are our strengths, weaknesses, opportunities, and threats?
  What assets should we build on? What challenges must we overcome?
- At its ideal, what would our organization be like?
- To bring about our vision, what specific outcomes must we achieve?
- What forces will affect our efforts to achieve these goals, and how do
  we address them?
- What is the most effective approach to reaching those outcomes?
  What are the measures of success?
- Have we engaged the people and organizations that need to be
  involved?
- How do we ensure accountability to our organization and the
  community?
- Are our strategies moving us effectively and efficiently toward our
  goals? How much progress have we made so far? What else needs to
  be done?

You will note that the survey addresses the activities that made up the
five phases of the Organizational Alignment Cycle. Based upon the results
of this survey, it may be necessary to go back to one of the previous phases
and make additional changes to prevent a problem or to get the maximum
effectiveness related to the Organizational Alignment process.

A Lean organization is one that has aligned it strategy, goals, and energy
focused upon a common vision.

—HJH

10

Epilogue

You just wasted a lot of your time unless you put the ideas to work.

—HJH

It is only fair to ask the question, "Is the Organization Alignment approach presented in this book the only one that will work? The answer is no. A technique called Policy Deployment has also been effective. We believe that it is best used with a small organization. The approach we presented is much more controlled and more comprehensive.

Throughout this book, the focus has been on helping organizations and their leadership to develop a process of alignment to help determine what needs to change in order to always remain a vital force in the marketplace. Organizational Alignment is a basic business discipline that addresses both the processes and behaviors in a very systemic way that drives positive results. Its objective is to create an organization in which the employees and management devote their time and energy toward accomplishing results without wasting effort overcoming obstacles.

Don't try to achieve complete alignment, for that requires a very stable environment that is undesirable in most organizations.

—HJH

One organization that we worked with that applied the Organizational Alignment methodology obtained the following results:

- Reduced two levels of management
- Reduced the management team size by 20%
- Improved customer satisfaction from 75% to 81%

Established a Knowledge Management System

Identified five processes that by outsourcing would reduce cost from 15% to 25% while obtaining higher quality output

Reduced cycle time and processing time on six processes

Strategic challenges will force management to change and innovate in the next decade, sometimes dramatically and sometimes incrementally. Although many of the changes have already begun, many more will be needed. Organizational Alignment can function as a change accelerator by developing new ways of managing change and stress, as well as improved information systems. Increasing the competition factor will demand improvements in competitor marketing intelligence systems, as well as new strategies for customer linkages.

> The current work environment is rich in social, psychological and political drivers that cause fatigue and eventually lead to burnout.
>
> **—Barbara Kaufman, President, ROI Consulting Group**

The globalization of business will require global strategies and structures, global cultures, and creative management styles. Changing technology will provide a new way of accelerating product life cycles and new product concepts to achieve new competitive advantages. A diverse workforce will require new leadership and management styles, new benefits, and new reward systems. The transition to a knowledge-based society will require new, innovative management paradigms and knowledge management techniques. Finally, increasing complexity will demand more expert systems and computer simulations, which will give leadership a prominent place in the Organizational Alignment process of managing and aligning resources and processes.

At the beginning of this book, we asked you to visualize yourself as the CEO of a Fortune 500 company where you have just left a board meeting at which the members of the board expressed their dissatisfaction with progress achieved last year. In your head you knew that there should be a rational explanation for the failure of the organization to change, and in your heart you felt that, even though there were insurmountable obstacles, the answers were just around the corner. By reading this book, you have turned the corner and have come up with a number of alignment strategies to save not only the organization but your job as well. And now is the hour to begin the implementation. The time is short, the need is great, and the future is yours.

Organizations would benefit highly from encouraging and helping to develop strategic thinking in as large a number of their employees as practicable.

—Iraj Tavakoli and Judith Lawton, "Strategic Thinking and Knowledge Management"*

Figure 10.1 is a set of summary recommendations.

Organizational Alignment begins and ends with strategy. The first challenge that we observed in this book was to be able to translate the organization's strategy into operational terms. Our discussions focused on the

	Goals	Structure	Management
Organization	*# 1* – Establish an alignment model and Balanced Scorecard (BSC), with goals that are aligned with the organization's mission.	*# 4* – Determine the size and scope of the extended organizational alignment and the organizational design (OD) of the "modules" involved.	*# 7* – Use a Business Performance Management (BPM) scorecard and assess employee, staff, and organizational needs. Evaluate results through cost-benefit analysis.
Process	*# 2* – Map and benchmark processes by identifying best practices for aligning organizational design, development, and delivery of programs, services, and products.	*# 5* – Use benchmarks to set process standards and develop policies and practices to support the new alignment and business performance; Use information technology to integrate and document information flows.	*# 8* – Evaluate processes to determine alignment potential and resource usage levels. Use feedback to improve process capability, efficiency, and quality.
Job/ Performer	*# 3* – Identify roles needed, responsibilities, and outputs. Tie performance goals to reward systems.	*# 6* – Develop work flow and performance feedback channels that support best practices and yield high performance levels.	*# 9* – Set employee performance standards, measure results, and use feedback to improve performance for reward/recognition.

FIGURE 10.1
Summary of Recommendations.

---

* Iraj Tavakoli, Judith Lawton, (2005) Strategic thinking and knowledge management, *Handbook of Business Strategy*, 6:1, 155

use of strategy maps and balanced scorecards to highlight targets and goals, in order to drive improvement, which accelerates business results. As these initiatives get rationalized within the business, accountability is assigned and enhanced.

In their seminal work on alignment, Robert S. Kaplan and David P. Norton summarized the five key management processes that they had previously identified as important for successful implementation.*

- Mobilization and orchestrating change through leadership
- Strategy translation using maps, scorecards, targets, and initiatives
- Organizational Alignment of corporate and business units, support units and external partners, and systems, processes, and the boards with strategy
- Employee motivation
- Governance through management reviews

The greatest gap between the "Hall of Fame" organizations and others occurs because the Hall of Famers usually are much better at aligning their corporate, business unit strategies. This leads us to the belief that alignment produces dramatic results.

> Whether a company is established or new, the challenge for both managers and executives is to create and manage the infrastructure that enables the company to be flexible and agile.
>
> **—John Nisbet, founder and CEO, Kilcreggan Enterprises**

We covered the alignment of the organization to the strategy in terms of the organization's role. We moved the focus on aligning the Strategic Plan down from the executive office to the business units, and then embedding it into the support units and the entire supply value chain. Of critical importance is the alignment of the board of directors, which is often overlooked or minimized. In later chapters, through the use of Organizational Change Management and Area Activity Analysis, we covered the alignment of the human capital, so that the entire workforce was motivated to make the Organizational Alignment project a success. A most difficult challenge indeed!

---

* Robert S. Kaplan and David P. Norton *Alignment: Using the Balanced Scorecard to Create Corporate Synergies* (Boston: Harvard Business School Press, 2005).

Also, the alignment of planning and control systems was highlighted throughout the book, as well as the need to govern in such a manner as to make strategy a continual process, part of the organizational DNA. The need for interactive planning processes was stressed, in order to initiate and align planning activities. At the heart of these matters is the business of integrating IT and HR planning into the mix, and to provide for clearly aligned budget linkages. Operations management must include process improvement initiatives, which are aligned and enhanced through formal knowledge sharing and communication efforts.

Finally, the areas of learning and control were emphasized throughout the book, through the use of assessments, open-ended feedback, and balanced reporting systems. Management reviews are held frequently to review progress and assure tight alignment to the organizational Key Performance Indicators (KPIs). If you are an effective leader within your organization, it is your responsibility to mobilize change and create energy within your organization. You must have a strong commitment and be ready to clearly articulate your case for change. Using the Organizational Alignment principles and methods outlined in this book, you are now ready to engage your leadership team and clarify for vision and strategy. The organizational cataracts will begin to fall away as you outline a new way of managing and aligning your parts of the organization with the vision and strategy. The time is right; the need is great; the choice is yours.

> You are part of the change parade. You can be a bandleader or you can be the one that sweeps up the horse droppings after the parade has passed. It is up to you.
>
> —HJH

Appendix A—Definitions and Abbreviations

The Annual Operating Plan (AOP): is a formal statement of business short-range goals, the reasons they are believed to be attainable, the plans for reaching these goals, and the funding approved for each part of the organization (budget). It includes the implementation plan for the coming years (years 1–3) of the Strategic Plan. It may also contain background information about the organization or teams attempting to reach these goals. One of the end results are performance plans for each manager and employee who will be implementing the plan over the coming year. The Annual Operating Plan is often referred to as the Operating Plan (OP).

Authority: is the granting of the position holder certain rights, including the right to give direction to others and the right to reward and punish. These rights are called positional powers, and they belong to the position rather than to the position holder.

Belief: is an acceptance of something as being true. In fact, it may or may not be true.

Budgets: fund the departmental activities, are used to estimate and plan expenses, and provide the resources required to implement the tactics.

Business: is the utilization of capital as performance solutions to produce value in the form of results.

Business Plan: is a formal statement of a set of business goals, the reasons why they are believed attainable, and the plan for reaching those goals. It also contains background information about the organization or team attempting to reach those goals.

Business Process Improvement (BPI): covers the breakthrough improvement approaches Process Redesign, Process Reengineering, and Process Benchmarking.

Communication Diagram: is a pictorial view of the way information is transferred within an organization.

Core Capabilities: are the business processes that visibly provide value to the customer (e.g., Honda's dealer management processes).

Core Competencies: are the technologies and production skills that underlie an organization's products or services (e.g., Sony's skill at miniaturization).

Critical Success Factors: are the key things that the organization must do exceptionally well to overcome today's problems and the roadblocks to meeting the vision statements.

Customer-Centric Organizations: are organizations in which offerings are combined and integrated advice, services, and/or software in support of their products/services that result in customized offerings. They are driven by customer portfolios.

Customer-Focused Organizations: use extensive market research in defining their product offerings and design. They build a demand for the products they are able to produce. They invest in providing their frontline employees with the knowledge, products, and tools to provide effective and consistent customer service.

Department: is made up of a manager and the people who report directly to him or her. It covers all levels of management: top, middle, and line.

Division of Labor: defines the distribution of responsibility (while hierarchy defines the distribution of authority). Division of labor concerns the way jobs are grouped into organizational units.

Executive Team: is a management team, which usually includes the chief executive officer (CEO), chief operating officer (COO), and vice presidents of an organization. In some organizations senior key staff people are also included.

External Supplier: is a supplier that is not part of the customer's organizational structure.

Fast Analysis Solution Technique (FAST): is a breakthrough approach that focuses a group's attention on a subprocess for a 1- or 2-day meeting to define how the group can improve the process over the next 90 days. Before the end of the meeting, management approves or rejects the proposed improvements.

Five S (5S): created by Ohno Shingo, 5S is a workplace organization methodology that uses a list of five Japanese words: seiri, seiton, seiso, seiketsu, and shitsuke. Transliterated into English, they all start with the letter "S." They are (1) Sorting (seiri), (2) Straightening or setting in order (seiton), (3) Sweeping or shining or cleanliness/systematic cleaning (seiso), (4) Standardizing (seiketsu), and (5) Sustaining the discipline or self-discipline (shitsuke).

Flowchart: is a method of graphically describing a process (existing or proposed) by simple symbols and words to display the sequence of activities in the process.

Function: is an organization that is made up of a specific skill set of people (i.e., quality assurance, finance, human relations, production control, sales, marketing, product engineering, manufacturing, industrial engineering, etc.) or the total group of people that reports directly to a vice president of the organization.

Hierarchy: reflects the distribution of authority among organizational positions. It defines formal reporting relationships that map out upward communication channels through which management expects information to flow.

High-Impact Teams (HITs): is a breakthrough approach that focuses a group's attention on the processes that are going on within a specific area. It realigns the work area to minimize the movement of output between activities resulting in decreasing stock and shorter cycle time. A typical HIT activity will last for 2 weeks and between 70% to 80% of the future-state solutions will be implemented within the 2-week time period.

Individual Performance Indicators (IPIs): is a set of measurements that are used to measure the performance of individual managers and employees. These are key inputs to their performance plan and evaluation. Often they are included in the individual's job description.

Internal Supplier: is an area within an organizational structure that provides input into another area within the same organizational structure.

Key Business Drivers (KBD) (also called Controllable Factors): are the things within the organization that management can change that control and/or influence the organization's culture and/or the way the organization operates.

Key Business Drivers (KBD) Vision Statements (also called Controllable Factors Vision Statements): are statements of how the Key Business Drivers will be operating 5 years in the future. They are developed during the Strategic Improvement Planning process. There are usually 8 to 12 of these vision statements.

Key Business Driver Improvement Plans: are plans designed to transform the organization from the as-is state so that it is in line with the relevant KBD vision statement. They are usually 3- to 5-year plans.

Key Performance Indicators (KPI): is a set of measurements, usually no more than 10, which are used to evaluate the progress and performance of a total organization.

Long-Term Vision Statement: is usually prepared by top management and defines the state of an organization 10 to 25 years in the future from the date that the statement was created.

Mission: is the stated reason for the existence of the organization. It is usually prepared by the CEO and seldom changes, normally only when the organization decides to pursue a completely new market.

Natural Work Team: is any group of people that consists of a manager or a team leader and the people who report directly to him or her.

Organizational Alignment: is the methodology that brings the organization's structure, processes, networks, people, and reward system in harmony with the Strategic Business Plan and the Strategic Improvement Plan.

Organizational Master Plan (OMP): is the combination and alignment of an organization's Business Plan, Strategic Business Plan, Strategic Improvement Plan, Strategic Plan, and Annual Operating Plan.

Organizational Objectives: are used to define what the organization wishes to accomplish over the next 5 to 10 years. These are usually defined by top management.

Organization Segments: are smaller subgroups comprising like or supporting types of activities. They may be divided by geographic, industrial/market segment or user client needs.

Outcomes: are the measured results that the organization realized as a result of the action taken.

Performance Goals: quantify the results that will be obtained if the organizational objectives are reached.

Performance Plans: are contracts between management and the employees that define the employees' roles in accomplishing the tactics, and the budget limitations that the employees have placed upon them.

Planning Assumptions: are the conditions and inputs that the individual, who is doing the planning, will consider as part of his or her estimation.

Planning Ground Rules: are the conditions and the requirements that the plan is based upon. The organization's current conditions are a key input to the planning ground rules. The requirements include setting targets for the organization's KPIs.

Policy Deployment (also called Hoshin Kanri): is a method devised to capture and cement strategic goals, as well as flashes of insight about the future, and develop a means to bring these into reality. It is based upon Shewhart's Plan-Do-Check-Act cycle. It creates goals, selects control points, and links daily control activities to the organization's strategy.

Principle: is the ultimate source, a fundamental truth, and the motivating source upon which others are based. If a statement is ever not true or not followed, it is not a principle. (Note: Any manager or employee who is not living up to the organization's principles is a good candidate for separation from the organization.)

Process Map: is a hierarchical method for displaying processes that illustrates how a product or transaction is processed. It is a visual representation of the work flow either within a process or an image of the whole operation. Process mapping comprises a stream of activities that transform a well-defined input or set of inputs into a predefined set of outputs. It is a flowchart with inputs and outputs added to each activity, thereby increasing its value in refining processes.

Product-Centric Organizations: are organizations that have multiple product lines that divide into separate business lines and/or models. Often there are few or no interrelationships between the product lines (e.g., a computer manufacturer that also sells management consultant services). They are driven by product portfolios.

Short-Term Vision Statement: is usually prepared by top management and defines the desired state of an organization 5 or 10 years in the future from the date the statement was created.

Strategic Business Plan (SBP): focuses on what the organization is going to do to grow its market. It is designed to answer the following questions: What do we do? How can we beat or avoid the competition? It is directed at the product and/or services that the organization provides as viewed by the outside world. When an organization is just being funded, this plan is often called simply a Business Plan.

Strategic Excellence Positions: are unique and distinctive capabilities that are valued by the customer and provide a basis for competitive advantage (e.g., Avon's distribution system).

Strategic Focus: comprises the key things that will set the organization apart from its competitors over the next 5 years. This list is defined by top- and middle-level managers.

Strategic Improvement Plan (SIP): focuses on how to change the culture of the organization. It is designed to answer the following questions: How do we excel? How can we increase value to all the stakeholders? It addresses how the controllable factors within the organization can be changed to improve the organization's reputation and performance.

Strategic Management (SM): is the process of specifying the organization's mission, vision, and objectives; developing policies and plans, often in terms of projects and programs that are designed to achieve these objectives; and then allocating resources to implement policies and plans, projects and programs.

Strategic Plan: is a document that is the result of Strategic Planning. It documents the organization's strategy and/or direction and makes decisions on the allocation of resources in pursuit of the organization's strategy, including its capital and people. It focuses on the future of the organization and is the combination of the Strategic Business Plan and the Strategic Improvement Plan, with each item prioritized to maximize the organization's performance.

Strategic Planning: is the systematic and more or less formalized effort of an organization to establish basic organizational purposes, objectives, policies, and strategies and to develop detailed plans to implement policies and strategies to achieve objectives and basic organizational purposes.

Strategies: define the approaches that will be used to meet the performance goals.

Supplier: is an organization that provides a product (input) to the customer (source ISO 8402).

Supply Chain Management: is the creation of a management process for integrating decisions, plans, and information systems from customer requirements through the fulfillment process to the suppliers of materials.

Tactics: define how the strategies will be implemented. They explain how the strategies will be accomplished.

Transition: is defined as an orderly passage from one state, condition, or action to another.

Value Statements: are documented directives that set behavioral patterns for all employees. They are the basic beliefs that the organization is founded upon, the principles that make up the organization's culture. They are deeply engrained operating rules or guiding principles of an organization that should not be compromised. They are rarely changed. Value statements are sometimes called operating principles, guiding principles, basic beliefs, or operating rules.

Vision: is a documented or mental description or picture of a desired future state of an organization, process, team, KBD, or activity.

Appendix B—Some of the 1100 Plus Improvement Tools

The following is a list of some of the 1100 plus improvement tools and methodologies:

7-S Model (Organizational Effectiveness)
Acceptable Process Level
Acceptable Quality Level
Acceptance Control Charts
Acceptance Sampling
Acquisition Streamlining
Action Diagramming
Action Planning
Activity Accounting
Activity Analysis
Activity Based Costing
Activity Cost Pool Definition
Affinity Diagrams
Analysis and Segmentation of Customer Views
Analysis of Customer Wants
Analysis of Variance/ANOVA
Annual Strategic Quality Plans
Application Construction
Application Installation
Application Structuring and Identification
Application Testing
Area Activity Analysis (AAA)
Association Diagramming
Assumptions Evaluation
Attribute Acceptance Sampling
Attribute Control Chart
Attribute Data
Attribute Identification
Attribute Measurement

Attribute Sampling
Attribute Sampling Procedures
Attribute Sampling Tables
Auditing
Audits by Top Management
Automatic Test Equipment
Automation
Autonomation
Average
Average (x) Chart
Awards
Baldrige Award
Barplots
Batch Procedure Design
Bayes' Theorem
Benchmarking
Benefits Assessment
Big-Picture Analysis
Black Belt Training
Block (Random) Sampling
Box Plots
BPI Measurement Methods
Brainstorming
Bureaucracy Elimination Methods
Business Area Data Modeling
Business Performance Management (BPM), BPM Measurement
Business Strategy Analysis
Business Systems Planning
Business Transaction Identification
Capacity and Staff Planning
Career Development
Career Planning
Cash Bonuses
Cause-and-Effect Diagrams (Fishbone Diagrams)
Cause-and-Effect Diagrams with Cards (CEDAC)
C Charts
Central Tendency Measurement
Chain Sampling Plans
Charts

Checklists
Checksheet Design
Chi Square Distribution/Test
Classification of Characteristics
Clearance Fits
Clearing Interval
Code Generation
Competitive Analysis
Competitor Product Disassembly Research
Computers
Computer Simulation
Concurrent Engineering
Conditional Probability
Consensus Building
Consumer Risk Quality (CRQ)
Context Diagramming
Continuous Sampling Plan
Contract Negotiation
Control Array
Control Charts; P Charts; u Charts; X Bar and R Charts
Control Limits
Control of Quality
Corrective Action
Correlation Analysis
Cost-Benefit Analysis
Cost-Cycle Time Analysis
Cost-Driven Analysis
Cost-Effectiveness Programs
Cost Flow Diagramming
Cost of Non-conformances
Cost of (Poor) Quality (COPQ)
Cost-Time Analysis
Cost-Time Charts
Could Cost
Counting Rules
Creative Brainstorming
Critical Success Factor Analysis
Criticality Analysis
Critical-to-Quality (CTQ)

Cross-Functional Management
Crossplots
Cumulative Hazard Sheet
Cumulative Sum Control Chart
Current Systems Investigation
Customer Analysis
Customer Data Analysis and Action Plans
Customer Interface Training
Customer Loss Analysis (Cost Impact)
Customer Partnerships
Customer Phone Calls (Management and Employees)
Customer-Related Measurements
Customer Requirements Mapping
Customer Reviews
Customer Round Tables
Customer Satisfaction Standards/Ratings
Customer Simulated Testing
Customer Surveys
Customer Visits
CUSUM Chart
Cycle Time Reporting
Cycle-Time Reduction Methods
Data Access Modeling
Database Generation
Data Collection
Data Flow Diagramming
Data Gathering by Document Review
Data Gathering by Interview
Data Gathering by Samples and Surveys
Data Gathering by Secondary Research
Data Sheets
Data Stratification
Data Structure Diagramming
Data View Identification
Decision Limits
Delphi Methods
Delphi Narrowing Technique
Department Improvement Teams (DITs)
Design of Experiments

Design Quality
Design Review
Design to Production Transition
Discreet Event Simulation
Discreet Probability Distributions

- Binomial
- Discreet Uniform
- Hypergeometric
- Multinomial
- Negative Binomial
- Poisson

Dispersion Measures
Distributions; see also normal distribution; probability distributions; reliability; chi square; etc.
Documentation Design
Dodge-Romig System
Double Sampling Plans
Double Specification Limit Plans
Education Cost Sharing
Education Design and Development
Effective Delegation
Effective Listening
Empirical Quantile Plots
Employee (Satisfaction) Surveys
Employee Training
Employee Visits
Environmental Pride Process (Your Living Rooms)
Error-Proofing
Error-Proofing Fixtures and Methods
Established Reliability
Estimation
European Quality Award
Evolutionary Operation (EVOP)
Executive Improvement Teams (EITs)
Executive Information Needs Analysis
Executive Needs Analysis
Exemplary Facilities

Expected Values

Exponential Formula for Reliability; Exponentially Weighted Moving Average; EWMA Chart

Facilitators

Facilitator Training

Facilities Planning

Failure Analysis

Failure Mode Effects Analysis (FMEA)

Fault Tree Analysis

F Distributions

Fewer Good Suppliers

Field Reporting

Financial Analysis

Financial Reporting

Five Whys

Flattening the Organization

Flowcharting

FOCUS

Focus Groups

Foolproof Engineering Methods

Force Field Analysis

Forms Design

Fractional Acceptance Numbers Plan

Frequency Distributions

Function Diagrams

Gain Sharing

General Ledger Analysis and Consolidation

Geometric Distributions

Goal Control Process

Good Manufacturing Practices (GMPs)

Graphical Methods

Group Presentation

Group Recognition

Group Sequential Plans

Harrington's Ten Fundamental Tools

- Bureaucracy
- Business Process Improvement Concepts
- Cost and Cycle Time Estimating (Activity Based Costing)

- Customer Needs/Expectation Assessment
- Flowcharting
- Measurement Methods
- Non-Value-Added Activities Elimination
- Process Simplification
- Simple English
- Walk-Throughs

Harrington's Ten Sophisticated Tools

- Benchmarking
- Business Systems Planning (BSP)
- Information Engineering
- PERT Charting
- Poor-Quality Cost
- Process Analysis Technique (PAT)
- Quality Function Deployment (QFD)
- Structured Analysis/Design (SASD)
- Value Analysis
- Value Control

Hawthorne Effect
Histograms
Historical Analysis
Hoshin Planning
Human Factors Engineering
Hypergeometric Distribution
Hypothesis Testing
Improvement Process Measurements
Indifference Quality Level
Individual Private Recognition
Individual Public Recognition
Individuals (x) Chart
Industrial Modernization Incentives Program (IMIP)
Inference Testing: see Statistical Inference
Information Engineering
In-Process Data Collection
Inspection and Testing
Inspection Planning

Instruments
Interference Fits
Internal Customer Concepts
Internal Customer Measurements
Interviewing Techniques
Investment Management
Ishiwaka Diagram
ISO-9000 Compliance; ISO-2959-1; ISO-3534-2; ISO-3951; ISO-8402; ISO-19011
Job Descriptions
Job-Related Training
Job Training and Certification
Just-In-Time
JUSE
Leadership Skills Development

- Coaching
- Communication
- Inspiring
- Listening
- Modeling

Kano Method; Total Quality Creation
Kolmogorov-Smirnov Test
Kume Quality Rules
Leadership Skills Training
Life Cycle Costing
Limiting Quality; Limiting Quality Level
Line Management
Location Plots
Logical Database Design
Lognormal Distribution
Long-Range Quality Planning
Loss Function; also see Taguchi Loss Function
Lot Plot
Lot Tolerance Percent Defective (LTPD)
LQL; see Limiting Quality Level
Maintainability Analysis
Maintainability Assessment

Management by Objectives
Management by Objectives Achievement Measurement
Management by Walking Around
Management Improvement Teams (MITs)
Management Presentations
Management Self Audits
Management's Seven Tools

- Affinity Program (K-J)
- Arrow Diagram
- Interrelationship Diagrams
- Matrix Charts
- Matrix Data Analysis
- Process Decision Program Chart
- Tree Diagram

Manufacturability Assessment
Market Analysis
Matched Plans, Derivation of
Maximum Standard Deviation
Mean Time Between Failures (MTBF)
Measurement; Central Tendency, Robust Design
Median Range
Median x Chart
Meeting Management
Metrology
MIL-Q-9858A Compliance; MIL-HDBK-217,338; MIL-STD-
   105D,105E,756,756B,781,785, 1472, 1543, 1629/35
Mind Maps
Mission Statements
Modified Acceptance Control Charts; Modified Control Charts;
   Modified Control Limits
Monetary Awards
Monte Carlo Sampling
Moving Range
Multiple-Attribute Decision Modeling (MADM)
Multiple Sampling Plan
Multi-Skills Maintenance
Multi-Skills Operator

Multi-Vari Strategy

Multivariable Analysis; Multivariate Analysis

Narrow-Limit Gauging

National Association of Testing Authorities (NATA)

Negative Analysis

New Employee Selection

New Performance Standards

No-Value-Added Analysis

Noise Plan; Noise Variables; Nomograph Continuous Sampling Plans (CSP)

Non-Conformances; Non-Conforming Spacing

Nominal Group Techniques

Nondestructive Testing and Training

Nonverbal Communications

Normal Distribution

Notched Box Plots

np Charts

On-Line Conversion Design

Operating Characteristic (OC) Curve

Operational Alternatives Analysis

Operational Definitions

Operation Verification

Organization Design

Organizational Analysis

Organizational Change Management Techniques

Organizational/Human Resource Enablement Sessions

Orthogonal Polynomial

Package Software Evaluation

Package Validation Testing

Paperwork Simplification Techniques

Pareto Diagram; also see the 80/20 Rule = the Pareto Principle; the Pareto Pyramid

Participative Management

Pay for Knowledge System

Pay for Performance

Pearson Coefficient of Correlation

Performance Planning and Evaluation

Personality Profile (Keirsey-Bates)

Physical Database Design

PIC-A-Solution
Plan, Do, Check, Act Cycle
Planned Experimentation, robust design
Poisson Distribution
Policy Deployment; also see Hoshin Kanri
Poor-Quality Cost; also see Cost of Quality
PRE—Control
Principal Component
Principal Component Analysis
Prioritization Matrixes
Prioritization through Ratings
Probability Distributions
Problem Solving
Problem Tracking Logs
Procedure Identification and Specification
Process Analysis and Improvement
Process Analysis Techniques
Process Capability Analysis
Process Capability Studies
Process Control Techniques
Process Decision Program Chart
Process Documentation
Process Engineering
Process Flow Controls
Process Improvement Teams (PITs)
Process Modeling
Process Qualification
Process Performance and Capability
Process Simplification Techniques
Process Walk-Through Methods
Process Window Definitions
Producers Risk Quality (PRQ)
Product Cycle Controls
Product Design Assurance
Product Quality, Eight Dimensions of
Program Evaluation and Review Technique (PERT) Charting
Projection Analysis
Prototype Test Checklists
Prototyping

Quality, definition of; auditors; control of; design vs. manufactured; specifications/requirements for

Quality Area Improvement

Quality Assurance Planning

Quality Characteristics

Quality Communication

Quality Company Policies

Quality Control, also see Control of Quality

Quality Control Circles (QCCs)

Quality Data Collection and Reporting System

Quality Engineering Methods and Training

Quality Function Deployment

Quality Improvement; also Quality Improvement Teams (QITs)

Quality Integration

Quality Loss Function

Quality Manuals

Quality Policy

Quality Policy Deployment

Quality Systems

Quality Visions

Quantile-Quantile Plots

Quincunx; Quincunx Data

Random (Block) Sampling

Range (R) Chart

Rejectable Process Level (RPL)

Reject Control Charts

Regression Analysis

Reliability Analysis

Reliability-Centered Maintenance

Reliability Predictions

Report Design

Requests for Corrective Action (RCA)

Requirements; also see Customer Requirements

Resource and Activity Driver Analysis

Responsibility Charting

Restructuring of the Quality Assurance Organization

Reverse Thinking

Risk Assessment

Risk/Opportunity Management Process

Risk Taking
Robust Design
Rules of Probability, Combinatorics
Sample; Sample Averages/Range; Sample Plan; Sampling System/Scheme
Sample Inspection
Sampling Techniques; also see Acceptance Sampling
SCAMPER
Scatter Diagram; Scatter Plots
Screen Design
Security and Access Control Design
Self-Control; Process Control Techniques
Self-Managed Work Teams
Sequential Analysis
Sequential Sampling Plans
Set-Up Time Reduction
Seven Basic Tools

- Brainstorming
- Cause and Effect Diagrams
- Check Sheets
- Data Stratification
- Histograms
- Pareto Diagrams
- Scatter Diagrams

Ship-to-Stock Cost
Simple English
Simple Language Analysis
Single Specification Limit Plans
Six Sigma; System of Measurement and Problem Solving
Skip Lot Inspection; Skip-Lot Sampling Plans
Software Quality Assurance
Solution Analysis Diagrams
Solutions Evaluation
Source Inspection
Specification; Specification Limits and Tolerances
Stakeholder Needs Analysis
Standards
Standard Deviation; Charts/Method

Standardization
Statistical Design of Experiments

- Factorial Design
- Fractional Factorial Experiments
- Grace Latin Square
- Latin Square Model
- Orthogonal Arrays

Statistical Estimation

- Bayesian Estimates
- Confidence Intervals
- Prediction Intervals
- Statistical Inference
- Tolerance Intervals

Statistical Methods (Control Charts)
Statistical Process Control

- C Charts
- np Charts
- p Charts
- u Charts
- X Bar and R Charts

Statistical Thinking
Statistical Tolerance Intervals/Levels/Limits
Stem and Leaf Plots
Stock Purchase Plans
Storyboarding
Strategic Business Review
Strategic Planning Process
Stress Testing
Stress Testing (Test to Failure)
Structural Customer Surveys
Structural Methodology for Process Improvement
Structure Charts
Structured Analysis

Structured Customer Surveys
Structured Design
Suggestion Programs
Supplier Design Involvement
Supplier Partnerships
Supplier Process Audits
Supplier Qualification
Supplier Quality Incentive Plans
Supplier Ratings
Supplier Seminars
Supplier Surveys
System Reliability
Systematic Sampling Plan
Systems Assurance
Taguchi Techniques
Target Costing
Target Goal Setting
Task Team (TT)
Team Building
Team Recognition
Teams-Group Process
Technology Enablement Sessions
Technology Impact Analysis
Test Objectives Definition; Test Equipment
Test of Hypotheses
Test Plan Design
Theoretical Quantile-Quantile Plots
Three-Dimensional Bar Plots
Time Management
Tolerance Limits; Tolerances
Total Quality Management (TQM)
Total Strategic Quality
Training Programs
Type I, II, III Conflicts
V-Mask
Value Analysis
Value Analysis Engineering
Value Engineering
Variability

Variable Control Charts
Variables Data
Variation
Visioning; also Vision Statements
Visual Controls
White-Collar Production Management
Wiebull Chart
Work Cells
Work Flow Analysis
Work Simplification
Work Teams
X Charts
Zero Defects
Zero Stock (Just-In-Time)

Index

For Product Safety Concerns and Information please contact our EU
representative GPSR@taylorandfrancis.com Taylor & Francis Verlag GmbH,
Kaufingerstraße 24, 80331 München, Germany

Printed and bound by CPI Group (UK) Ltd, Croydon, CR0 4YY
08/05/2025
01864371-0002